Napier,

You Kibitzed so faithfully that I consider you a collaborator in this venture.

Thanks, Les

NATIONAL OIL COMPANIES

NATIONAL OIL COMPANIES

Leslie E. Grayson
University of Virginia

JOHN WILEY AND SONS

Chichester · New York · Brisbane · Toronto

British Library Cataloguing in Publication Data:
Grayson, L. E.
National oil companies.
1. Petroleum industry and trade–Europe
I. Title
338.2'7'282094 HD9575 80-41436

ISBN 0 471 27861 0

Typeset by Pintail Studios Ltd., Ringwood, Hampshire
Printed in Great Britain by Page Bros. (Norwich) Ltd., Norwich.

Acknowledgments

Two institutions and many individuals made this study possible. The Ford Foundation provided financial assistance for a year and a half (1975–77) that allowed me to make a serious start on the book. I owe the most gratitude to Peter de Janosi, who urged me to 'get on with it.' My thanks also go to Marshall Robinson. The friendship of these men (now both with the Russell Sage Foundation), in addition to the Ford Foundation's support, helped me to carry out the project.

The other institution that played a critical role in the study is INSEAD (the European Institute of Business Administration) in Fontainebleau, France. INSEAD proved to be ideal, both geographically and intellectually, for conducting this study. The rest of Europe was easily accessible and Fontainebleau was a beautiful place to live. The international character of INSEAD's faculty and student body created an atmosphere in which I thrived. Guy de Carmoy helped me to conceptualize the project. My very special debt is to Martin Flash, who assumed major responsibility for Chapters 2, 4, and 5 with imagination, diligence, and, above all, with a sense of friendship. Mary Lou Carter ably assisted me with Chapter 3; John Cutts with Chapter 6; Richard Paniguian with Chapter 7; and David Lilley with Chapter 8. Veronique LeCocq, Janice Selkirk, and Marie Wright typed numerous drafts of Chapters 2–8 under sometimes difficult conditions yet kept their sense of humor (at least most of the time). If there are any remaining errors, the blame must stay with me.

My editor at Wiley, James Cameron, and his readers responded quickly to drafts that were not always on schedule. At the University of Virginia, Lonnie Beagle typed Chapters 1 and 9–13 expeditiously and with good cheer. The same chapters were edited by Beverly Seng with results that only I can fully appreciate.

I had the anonymous cooperation of executives of national oil companies and of the affiliates of most multinational oil companies in France, Italy, Germany, the United Kingdom, and Norway. In the same five countries I also enjoyed the confidence of government officials. In addition, the appropriate international organizations proved to be eager to assist. Altogether, I interviewed approximately 120 executives and officials, and these interviews provided invaluable raw material for the study.

Our two daughters, Carol and Judy, were proud that their father was engaged in an enterprise that they considered important. My wife Olivia has always supported my work; her help was most substantial with this particular study. I dedicate this book to her.

Acknowledgments

Contents

Contents

Introduction

In 1970, about 70 per cent of the world's oil trade was handled by the oil multinationals – Exxon, Royal Dutch/Shell, Mobil, Texaco, Standard of California, Gulf, and British Petroleum. A decade later, the multinationals' share has declined to about 50 per cent. Some of the trade no longer handled by the MNCs has moved to the 'spot' market in which both the multinationals and the national oil companies participate. A greater part, however, has shifted to markets served by the national oil companies of the producing and consuming countries.

But despite their increasing importance, the national oil companies are not well understood. Compared to American companies, European companies provide little information about themselves, and European public enterprises offer even less. Little research has been done on the European national oil companies, and much of what has been published is anecdotal. Moreover, the published materials usually focus on individual companies and therefore lack the perspective that only a comprehensive study of the national oil companies could provide. This book seeks to remedy both problems. It offers analytic case studies of each of the six major national oil companies, and, in several synthesizing chapters, it studies the entire group as an important emerging phenomenon of the international energy scene.

The six companies included in my study are Compagnie Française des Pétroles (CFP) and Société National Elf-Aquitaine (SNEA) in France, ENI in Italy, Veba in Germany, British National Oil Company (BNOC) in the United Kingdom, and STATOIL in Norway. I excluded several others because they are not as large as the ones I dealt with here and also because of the practical difficulties of such an extended study. Hence, excluded are OMV of Austria, Neste of Finland, Swedish Petroleum of Sweden, Petrogal of Portugal, and CEPSA, ENPETROL, CAMPSA, and HISPANOIL of Spain. I excluded British Petroleum and Petrofina because they always operated as private enterprises; the same reason applies to Burmah Oil.

This book is special in its methodology, considering its subject matter. Because published material on the national oil companies is generally trivial and superficial, I supplemented my review of the available literature with field research. I interviewed some 120 executives of national and multinational oil companies, as well as government officials. Although I had been warned that executives and officials would not talk to outsiders, I found them immensely cooperative and willing to provide both facts and viewpoints. I attribute their cooperation to their interest in the project, to the new posture of 'transparency' in the oil industry, and to the fact that, having been myself in the oil industry for twelve years, I was not regarded as a neophyte. The interviews took me to Paris, London, the Hague, Brussels, Bonn, Cologne, Dusseldorf, Hamburg, Geneva, Milan, Rome, Oslo, Stavanger, New York, and Washington.

Moreover, the executives and officials who cooperated with me on the study graciously agreed to review my drafts and did, in fact, comment on them extensively. Thus, I was doubly fortunate: their help enabled me to get the facts straight, yet I retained the freedom to express my own judgments.

The six analytic chapters on individual companies discuss their managerial structures as well as their performance within their immediate economic and political environments. Thus, these chapters describe the national oil companies' objectives, their strategies, their relationships with owner governments and with multinational oil companies, and, finally, their impact on the structure of the international oil industry.

The future strength and importance of the national oil companies will depend largely on their access to crude, their technological contributions to the search for new sources, their profits and need for governmental financing, and the competition that they face for funds from national treasuries. National ideological trends may also alter these companies' relationships with their owner governments. However, the effects of political philosophy on these companies can be exaggerated, as two recent examples have shown. In the spring of 1979, the Tories came to power in the United Kingdom. Both during and after the election, the Tories announced that BNOC's charter would be substantially curtailed, yet little of substance was actually changed. In the fall of 1979, the new Tory government in Canada proposed to 'denationalize' Petro-Canada, but aside from some restructuring, Petro-Canada remained the state oil corporation.

Six synthesizing chapters offer wider perspectives on the national oil companies by making comparisons among the six companies and between nationals and multinationals. For example, I was able to compare the management policies and techniques of the nationals to those of the multinationals. I also traced the evolution of managerial policies and techniques within the national oil companies, formulating a hypothesis concerning the managerial 'life cycle' within these companies.

These chapters also examine the national oil companies within their larger economic and political context – the international oil industry itself. Our traditional means of describing economic and business relationships are challenged by the phenomenon of the national oil company. What is the meaning of the term 'competition' in a changing business environment wherein state-owned companies compete with private firms? How do we define the 'markets' of a national company whose objectives are not solely to earn profits in oil trade but to perform certain politically determined economic and social functions? The term 'national' is itself ambiguous: is it the company or the oil that is state owned?

Finally, the book explores the larger issue of what is the most efficient and equitable relationship of the nation-state to a large and crucial multinational industry. As the Planning Director of ENI remarked: 'Oil is a political commodity now. It is not something to be left to markets and businessmen.'[1] While this is an overstatement – the multinationals as a group still handle a great deal more oil than do the national companies – this assessment is partly supported by trends of the seventies.

While my study deals with European national oil companies, it has relevance for the United States. Since the oil 'crisis' of 1973–74 and its renewal in 1979, politicians have sporadically discussed the establishment of a US national oil company. An early (1974) proposal was made in the US Congress by Senator Adlai Stevenson.[2] The predecessor organization to the present US Department of Energy obtained a major study on the federal government's options in dealing with world oil markets.[3] One of the alternatives presented was to create a petroleum corporation owned by the federal government.

In the wake of yet another dramatic rise in oil prices in 1979, the call for governmental involvement in the oil industry has been renewed. President Carter has proposed creating an Energy Security Corporation that would invest $88 billion in a ten-year period, its purpose being to produce 2.5 million barrels of synthetic fuel per day, largely from shale and coal.[4] The proposal is now under congressional consideration. The Department of Energy is also considering direct intervention – although only at the importing level.[5] Proposals range from federal scrutiny of imports made by private companies to federal takeover of all imports. The Department has not yet agreed on any of these proposals, nor has the President recommended a specific alternative. If such agreement is reached, the proposals will have to be passed by Congress. Whatever the fate of these proposals, it is clear that the issue of federal involvement continues to absorb US policymakers. The experience of Western European countries with their national oil companies ought to help policymakers evaluate America's own options.

References

1. Quoted in the *New York Times*, December 30, 1979.
2. 'A Federal Competitor in Oil,' *New York Times*, February 25, 1974.
3. 'An Evolution of the Options of the U.S. Government in Its Relationship to U.S. Firms in International Petroleum Affairs,' *Federal Energy News*, FEA, Washington, February 12, 1975.
4. 'A Desperate Search for Synthetic Fuels,' *Business Week*, July 30, 1979.
5. 'Energy Agency Is Drafting Plans for U.S. to Take Direct Role in Importing Crude,' *Wall Street Journal*, January 23, 1980.

CHAPTER 1

Why National Oil Companies?

Some have said, only partly in jest, that more write about the oil industry than work in it. The writers concentrate on the Organization of Petroleum Exporting Countries (OPEC) and on the multinational oil companies (oil MNCs), also known as the 'majors' or the 'Seven Sisters': Exxon, Royal-Dutch/Shell, Texaco, Mobil, Standard of California, Gulf, and British Petroleum.[1] In the major countries of Western Europe, however, the multinational oil company shares its dominion over the oil industry with the national oil company (NOC), a phenomenon largely neglected by writers.

The NOCs, owned or controlled by their respective governments, are large and influential, although little known and therefore not well understood. In the light of their significance, it is surprising that so little research has been done on them.[2] In the last fifteen years American 'independents' appeared, but they are now withdrawing from Europe, their operations largely taken over by the NOCs.[3]

This book seeks to remedy that situation by systematically focusing on the NOC as an important emerging factor on the Western European energy scene. The book deals with six national oil companies in five countries: in France, Compagnie Française des Pétroles (CFP) (established in 1924) and Société Nationale Elf-Aquitaine (SNEA) (1951 and 1966); in Italy, ENI (1926 and 1953); in Germany, Veba (1974); in Norway, STATOIL (1972); and in the United Kingdom, British National Oil Company (1976). Veba is the largest company in Germany (1977 net sales were $13.1 billion). CFP and SNEA are respectively the largest and third largest companies in France (1977 net sales $11.4 billion and $8.1 billion). ENI is the second largest company in Italy (1977 net sales $10.5 billion). Among industrial companies outside the United States, in terms of their 1977 net sales, Veba ranked 7, CFP ranked 10, and ENI ranked 14.[4]

The term 'national oil company' is ambiguous. This study defines as 'national' only those companies that have been used for national purposes. For instance, the British government holds 31 per cent of shares of British Petroleum and the Bank of England owns 20 per cent, yet the company has always been allowed to operate as a private enterprise. Petrofina is owned by private Belgian interests, and the Belgian government has never made any attempt to manage Petrofina. Since these two companies have not been used for national purposes, they have been excluded from the book. Also, it has never been made entirely clear as to what is 'national'. Is it the company, or the oil (and/or the gas)? Is the crude oil controlled by SNEA in Gabon 'national'? The presumptive answer to both of these questions has been yes – though it is far from clear that this answer is correct.

I have used two approaches in this analysis of the national oil companies. Case studies of the six companies comprise the body of the book – seven chapters. The first and concluding chapters synthesize the results of the case studies, raise some hypotheses, and test them empirically. Incorporating these two methods, this book examines the performance of the national oil companies, reviewing their strategies, evaluating their fulfillment of stated and implied objectives, and assessing their relationships to their home governments, to the multinational oil companies, and to the international structure of the oil industry.

History of Government Involvement in Oil

Energy is a 'commanding height' of any economy, and it has been a political issue ever since the first Royal Commission Report on the Coal Industry in Britain was issued in the middle of the nineteenth century.[5] Until recently, most people believed that while other energy industries were heavily regulated, the oil industry was unregulated. The former conception was certainly accurate. Governments have viewed coal as a natural resource of strategic importance, although the industry has been financially troubled since World War II. Natural gas and electricity have been regulated because they have to be transported by common carrier. Nuclear energy has been supported for both its technological and its military importance, and has been supervised to preserve the environment and the public safety.

Contrary to popular belief that the oil industry was an example *par excellence* of free enterprise, the oil industry has also been subject to governmental intervention for the past fifty years. True, the refining and marketing of petroleum products and the manufacture of petrochemicals have been largely free of state control or participation for a number of reasons. First, until the seventies, there was an abundance of competing suppliers. Second, since oil was cheap, its price was not of major economic importance. Also, logistics and technical factors allowed some degree of competition, and thus there was no real fear of monopolization. Finally, no major 'local' oil had been produced commercially in the large consuming countries of Western Europe until the North Sea oil came on stream.

Nevertheless, the industry has been heavily regulated since the twenties in Japan, France, and the United States (the US controlling only production and transportation). In Italy, the government has controlled the industry to the extent that post-World War II Italian governments are capable of exercising control. All major energy industries were therefore indeed regulated by the early seventies. In only two major OECD countries have oil companies acted freely in a 'market' economy – Germany and the United Kingdom. Actually, even after Veba and British National Oil Company (BNOC) were established, these two countries still have had relatively free oil markets.[6] One may, therefore, more correctly speak about an acceleration in intervention and not about new intervention in the oil industry.

The degree of intervention has varied from country to country and from time to time within each country. (The French are an exception; their policy has been consistent since 1928.) The degree of intervention has depended mainly on traditions of industrial policy (France and Italy have always had a strong government 'presence' in big business), on the degree of dependence on oil, and on whether or not the countries were 'hosts' to any of the oil MNCs. (Only the United States, the United Kingdom, and the Netherlands were host countries.) Most important, however, has been the consuming countries' varying dependence on oil. Of the five countries under review here, Germany and the United Kingdom had and still have significant portions of their energy supplied by coal; Norway's is supplied by hydroelectricity. France and Italy most rapidly became dependent on oil – all of it imported, mostly from OPEC countries.[7]

Thus among Western European governments, direct involvement in the oil industry preceded the oil crisis of 1973–74. It was, however, in the mid-seventies that countries realized that this crucial resource, having been available abundantly and inexpensively, had become relatively expensive and could become scarce. After 1973–74, energy became a paramount political issue and governments felt that their electorates had to see them taking some action. The 1973–74 oil embargo and the quadrupling of oil prices particularly shook France, Italy, and Germany, which were dependent on foreign MNCs. That the oil MNCs apparently allocated the scarce oil both efficiently and equitably did not really matter. What did concern these countries was that they could not control the decision making and perceived themselves to be helpless.

That the allocations during the embargo were equitable could be determined only after it ended. But during and immediately after the embargo, the policies of the oil MNCs were not easily intelligible either to European governments or to their publics. Lack of knowledge about the industry created the somewhat understandable apprehension that the oil MNCs might act to optimize either their own interests or those of their home governments, doing both at the expense of the continental European consumer.

Thus the embargo made European governments acutely aware of their lack of knowledge about the energy business. They resolved to rectify this situation by further direct participation, which would also enable them to react more effectively to any future crisis.

Each country requires optimal oil allocation, which may or may not overlap with optimal global allocation. No country is prepared to accept global allocation that clashes with its national objectives. The oil MNCs stand for optimal global allocation, and in significant parts of the world such allocation has coincided with individual nations' objectives. Where they did not – and for a host of reasons to be discussed in subsequent chapters, they did not entirely coincide in the five countries under review here – the governments had two options. One was to try to regulate the market indirectly via tariffs, taxes, and licenses, among other measures. The alternative was to intervene directly, to establish national oil companies.[8]

Why did these five countries choose the second option? Conventional economic analysts explain that national oil companies were formed to overcome the imperfections in international petroleum markets. However, the large number of entries in the early sixties indicate that these markets were more competitive than conventional analysts believed – the barriers to entry, while still significant, were lower than analysts assumed. Conventional economic analysis, then, offers scant explanation for the formation of national oil companies.

The thesis of this chapter, and indeed of this book, is that the national oil companies have been established to deal with specific economic, political, and social issues. These five countries found that indirect regulation through tariffs and taxes was not so much economically ineffective as politically and socially insufficient. In fact, direct intervention is more costly than indirect regulation. One must understand that NOCs cannot be launched, nor survive, nor prosper without consistent and sizeable governmental assistance. How long such assistance must be sustained depends on how rapidly the particular NOC manages to gain access to low-cost crude and significant market share in its home country.

Primary governmental efforts have been aimed at assisting NOCs to survive. The oil MNCs form an oligopoly that poses barriers to entry. These barriers are highest in production of crude, high in refining, and less high in marketing and transportation – except for pipeline transport, where barriers are very high. Barriers include the oligopolists' sheer size, their access to capital and to inexpensive and diversified crude supplies, and their worldwide marketing network. The worldwide acceleration of inflation since 1970 significantly exacerbated the penalty incurred by late entrants to the market. Before 1973, the NOCs had the entry problem nearly under control. But entry barriers were again raised when oil prices quadrupled and inflation climbed to double-digit proportions in Europe. The oil MNCs had much 'old' (i.e. pre-high-inflation) investment; the NOCs had to invest heavily in the post-inflationary environment. Moreover, since 1971, fiscal resources in most European countries have been increasingly constrained. The NOCs have had to compete vigorously for scarce public funds.

Despite these formidable costs, these five countries created national oil companies because their governments decided that the formulation and realization of their energy policies could not be left exclusively to the oil MNCs. The vagaries of international politics, finance, and trade were and are too uncertain to allow critical resources and industries dependent on them to be controlled by 'foreigners'. (This view was the amended version of the argument that oil is too important to leave in private, foreign, Anglo-Saxon hands – regardless of their past records.) National development strategies as well could not be dependent upon the judgment of outsiders – be those outsiders the OPEC or the oil MNCs. National oil companies were created for varying reasons, but most had three primary objectives: to reduce the state's dependence on the oil MNCs, to assure its domestic oil supplies, and to acquire knowledge of the oil industry. We shall consider each motive in turn.

Objectives of the National Oil Companies

Before outlining these objectives, I must caution that not all objectives apply equally to all six NOCs. Some are common to all, some apply to most, and some apply only to a single company. Each company's objectives are summarized in tabular form at the end of this section. At this point, we can roughly expect that the objectives of BNOC and STATOIL, both long on crude and having minimal or no downstream facilities, will differ from those of the four crude-short companies. CFP, in existence for more than half a century, will probably behave differently than do the newcomers. Veba, operating in a more or less *laissez faire* environment, will act more differently than do CFP, SNEA, and ENI, operating in more *dirigiste* settings.

The state's first objective was to reduce its dependence on the oil MNCs. Prior to the advent of the NOCs, the oil MNCs dominated Western European oil supplies by virtually controlling low-cost crude supplies as well as significant portions of transporting, refining, and marketing facilities. Large industrialized nations are loath to depend on foreign corporations' supply and control of the essential ingredients of economic growth. But this natural and generalized resentment was not the only reason that European countries planned to reduce their dependence on the MNCs. More specifically, these countries resented Anglo-Saxon domination; all of the MNCs were Anglo-Saxon, and five were American. Such nationalistic suspicion was hardly a new concern. In 1909, the British discovered oil in the Persian Gulf, and Churchill, then First Lord of the Admiralty, persuaded the British government to set up the first 'national' oil company, Anglo-Persian, the predecessor to British Petroleum. Churchill was afraid that the Royal Navy otherwise would be overcharged by Esso and Royal Dutch/Shell.[9] American domination has been very much resented by such an 'Anglo-Saxon' country as Canada, which established its NOC, Petro-Canada, in 1975. The President of Petro-Canada noted that one of his company's goals is to 'reassert Canadian control over more of the country's natural resources.'[10]

Western European countries also believed that diminishing their dependence on the MNCs would increase their own international political–economic powers. Countries with NOCs could better control their balances of payments and tax policies. NOCs could enable home countries to accumulate international business and technological expertise and could also be instrumental in foreign economic ventures, which would enhance the country's prestige at home and abroad. Thus by establishing NOCs, governments hoped to counterbalance the oil majors' powers and to augment their own.

The second key objective of national oil companies is to enable the home governments to develop the specific understanding of the oil industry needed to check the MNCs' activities. European countries have, of course, needed meaningful information about the oil industry ever since World War II. The need became paramount after 1973 when oil became sometimes scarce and always more expensive than it had been prior to the early seventies.[11] It is not entirely clear that governments must establish national oil companies to gain a 'window' on the industry; indirect means such as regulation might provide adequate information. Nonetheless, governments of France, Germany, Italy, the United Kingdom, and Norway judged that they indeed must participate directly to gain adequate knowledge about the industry.

This issue was recently aired in the United Kingdom. In 1973 a parliamentary committee noted that the oil MNCs were not then required to give the British government adequate cost data; the committee concluded that it was 'not satisfactory that the Energy Department should not have access to costs.'[12] The committee assumed that such information would enhance the government's control of oil and gas. The Conservative opposition argued that the current regulations provided adequate information. After extensive debate, the government did conclude that to get accurate information, the state must participate directly in oil industry operations. Resolution of this issue depends not on the quantity of information, but on the quality of its interpretation. Governments have been able to obtain – or could if they desired – virtually any information from the MNCs. But regulatory agencies in the five countries under review here did not have staffs qualified to analyze and interpret the data. Thus the NOCs aided their governments' understanding of the oil industry by interpreting oil industry information – whether that information was provided by themselves or by the MNCs.

Governments hoped that having a window on the oil industry would enable them to establish a 'yardstick' against which they could judge the performance of the oil MNCs in their countries. Alas, ruling a yardstick proved difficult. If NOCs were to serve as yardsticks, they would have to be similar to MNCs in all respects – namely, they could neither receive governmental assistance nor perform non-commercial services. But NOCs required such assistance and were often expected to perform services that hampered their financial performance. Hence they proved inappropriate measuring devices.

The third primary motive for creating national oil companies was to assure crude supplies that are inexpensive and reliable. Hopes of getting inexpensive oil have proven groundless, as we shall see below. Theoretically, national oil companies may be more reliable than MNCs for several reasons. First, during economic or political crises, MNCs may be pressured by their home governments to alter their allocation patterns – a disturbing possibility to European consumers. Neither is this concern new. France literally ran out of oil during the last months of World War I and was determined to obtain control over some of its supplies. On December 15, 1917, Clemenceau appealed to Wilson: 'The allies must not let France lack the petrol which is as necessary as blood in the battles of tomorrow.'[13] Because of this concern, in 1924 the French government obtained as part of its war reparation a quarter share of the Iraq Petroleum Company, originally destined for the Deutsche Bank. The state-owned Compagnie Française des Pétroles (CFP) was established on that occasion. Recent evidence suggests that political pressure has indeed been applied to the oil MNCs and that they have successfully resisted it.[14] It is their future ability to withstand these pressures that European governments doubt.

Second, national oil companies may gain preferential access to OPEC crude oil. OPEC countries, and/or their national oil companies, are thought to be more inclined to deal directly with consuming countries and their representatives – the NOCs – than with the oil MNCs. Thus, the OPEC countries may give the consuming countries preferential access to crude oil. This hope has also proven groundless, as we shall see below.

Third, national oil companies can cover abandoned markets and support weak ones. The oil MNCs have had and will augment a policy of market selectivity. They have withdrawn and will continue to withdraw from unprofitable markets, ignoring such markets' own interests. Furthermore, because of OPEC's economic power, the oil MNCs have become increasingly less profitable and thus unable to raise the vast investment funds needed to provide future energy needs. NOCs, backed by governmental guarantees, may be better able to raise such funds. Hence, these governments may insure their own oil markets – at a price, of course.

Finally, the NOCs might supplement the MNCs' exploration for long-term reserves as well as their production in fields whose yield might be increased by extra platforms. Because the social discount rate is very much lower than the private discount rate, investments that would be unattractive for an oil MNC might well be attractive for a NOC.

These triple objectives – reducing dependence on oil MNCs, gaining knowledge about the petroleum industry, and assuring oil supplies both at the crude oil and product market stage – remained crucial in international oil policy throughout the period up to the late seventies. They are likely to continue to be dominant considerations in the foreseeable future.

Once national oil companies were established, governments often set further domestic objectives aimed at increasing the governments' social, political, and economic control. Many of these domestic objectives changed over time and some of them are contradictory. Such discrepancies are to be expected among public ventures in an industry dominated by private companies. Some governments planned to use the NOC to allocate oil, an exhaustible and critical resource, for optimum public benefit. Many governments hoped to use the NOC to enact overall energy policies. The NOC then determined sufficient levels of investment in exploration and production, controlled rates of depletion, sought to optimize government income, and procured contracts and equipment orders. NOCs have located refineries and supply bases for exploration or marketing to bolster economically underdeveloped regions.

European governments also believed that backward integration into oil would protect their domestic petrochemical industries. These industries capture 'value added' within the countries' borders and thereby favorably affect their balances of payments.[15] Thus governments have encouraged their indigenous petrochemical industries, frequently protecting them with tariffs. Special chemical plants have been built as soon as petrochemical imports reached a level equivalent to the minimum efficient plant size. European governments feared that first the MNCs and, after 1971, OPEC might charge too much for crude oil or feedstocks. Consequently, these governments felt the need to have oil companies able to provide 'secure' supplies to their petrochemical companies.

Finally, most governments hoped to enjoy reasonable returns on their investments (ROIs). Discussing this objective last is not intended to imply that NOCs are less profitable than the oil MNCs; their profitability will be examined in subsequent chapters. Rather, I place this objective last to emphasize that if ROIs were

Table 1. Selected objectives of NOCs by companies

	'Window'	Reduce dependence on Anglo-Saxons and/or oil MNCs	Secure inexpensive oil	OPEC/NOC deals	Market selectivity, inadequate investment	Develop lagging regions/ instruments for foreign economic ventures	Managerial and technological expertise
CFP		X	X	X	?		X
SNEA	X	X	X	X	?	X	X
ENI	X	X	X	X	X	X	X
Veba		X	X	X			?
BNOC	X	X	NA	NA	NA	X	X
STATOIL	X	X	NA	NA	NA	X	X

Return on investment is excluded from this table because this objective is self-evident. NA = not applicable. ? = application of objective unclear.

the major or sole criterion, the oil industry would be left to private – and foreign – companies. NOCs were created to perform the sociopolitical–economic services just described because governments believed, justifiably or not, that these services cannot be adequately provided by the oil MNCs.

This book will examine the national oil companies' fulfillment of these various objectives: gaining independence from the MNCs, learning about the oil industry, securing oil supplies, and performing domestic political, social, and economic services. I will consider which of these objectives have been met, to what extent, how, and by which companies. The ensuing chapters will also weigh the social costs of implementing and of failing to implement the political, social, and economic services. In general, one must assume that the NOCs performed some of those services sufficiently satisfactorily to warrant their continued support. The NOCs under review here operate in democratic countries: France, Germany, Italy, the United Kingdom, and Norway. There seems to be no evidence of basic criticism of the existence of any of the NOCs under review. (Occasional criticism of their management policies should not be confused with challenging their existence.) Finally, this book will compare the national oil companies' performance with that of private companies. Comparisons with regard to return on investment, productivity, and costs will be presented in subsequent chapters. Such comparisons require careful interpretation, however, because private companies and NOCs have differing objectives; the several NOCs themselves have differing priorities.

Before such evaluations and comparisons can be made, it will be useful to explain in further detail some of the national oil companies' objectives.

Returns on Investment

Securing returns on investment and providing the social, political, and economic services described above do not conflict, theoretically. In practice, they often do.[16] In Sweden, for example, 'the largest state company controls the iron mines. It loses tremendous sums and it might be good management to close it up. But the government could not handle that politically, so the [holding] company gets used as a tool by whoever is in power.'[17] The greatest struggle for Air France is to 'attempt to put some distance between itself and the French government. Air France wants to make the government pay a direct subsidy to compensate for any loss-causing service it requires the airline to provide.'[18] These services include flying the Concorde to New York, Washington, and Rio de Janeiro; providing air links to some of France's overseas possessions; and serving both Paris airports, Orly and Charles de Gaulle. The government also forbids the airline to buy American jets. The strains inherent in government ownership – on the one hand the need to further national objectives through state intervention, and on the other the desire to make commercial sense of that intervention – have become apparent even in well-run state-owned corporations.[19] The chairman of ENI has been likened to a leader 'trying to run NASA at a profit with some responsibility for Indian Affairs and economic development, and he is working with a regulatory structure as backward as those you find in Latin America.'[20] The public wants to see the NOCs perform services and at the same time pay their way. This proves to be a tall order and, not surprisingly, either service or ROI falls by the wayside. For instance, when prices are kept relatively low by regulation, retained earnings available for future investment are reduced. Thereupon, either public borrowing or taxes have to be increased to provide funds for investment. Or, if the government directs the NOC to locate plants in areas that are not optimal from the NOC's point of view but are designed to economically bolster a country's less developed region, again added taxes or public borrowing must be provided to cover the value of the service performed. The viable firms in the oil industry are large and capital intensive. Thus the NOC's drain on public treasuries has sometimes been very great.

Governments tried to enable their NOCs to compete successfully with the oil MNCs by providing the NOCs access to inexpensive crude on the same basis as the oil MNCs, by restricting oil MNCs from increasing their marketing network and by imposing a variety of fiscal and regulatory subsidies. But one must also remember that ROI is not always considered top priority, neither by the government (which wants services performed) nor by the NOC itself. In fact, in certain cases low profits or losses are the only way to induce the government to supply additional capital for investment. If retaining independence from frequent government intervention is important to the NOC, it must maintain an adequate cash flow. An alternative NOC approach is not necessarily to maximize ROI but rather to minimize the need to go to parliament or treasury for additional funds.

Thus, although subsequent chapters include comparisons between public and private firms with regard to ROI, productivity (sales per employee), and the cost of crude oil delivered into major markets, such comparisons should be used with caution. Comparisons of ROI between NOCs should likewise be interpreted carefully. Even the most knowledgeable and impartial observers have differing views of NOCs' objectives, priorities, and changes in strategy; and some observers fail to perceive variations from company to company. 'Established' NOCs, such as CFP, maximize ROI; newly launched NOCs, such as STATOIL and BNOC, act (for the present at least) primarily as agents of the state. Even the two French companies behave differently, in response to their differing situation. SNEA claims to be *the* national oil company, while CFP sees itself as a French private oil company with a very significant government shareholding. In general, because of the number of nonfinancial objectives of the NOCs, their financial performance has been, almost by definition, poorer than those of the private companies. Their protests notwithstanding, the oil MNCs rather welcomed the NOCs for this very reason. To paraphrase Voltaire, if NOCs did not exist they would have to be invented, was no longer a joke by the late seventies.

Securing Crude Oil Supplies

One objective – a crucial one – has not been met by the established NOCs: the provision of inexpensive and reliable supplies of oil to the home country. Most oil is produced outside the main areas of consumption (the OECD countries) and consumed outside the principal production areas (the OPEC countries). There seems to be no evidence that the NOCs have been treated better than the oil MNCs in terms of either supply or price. In 1973–74 the search for oil by European governments and firms became frantic because of their vulnerability to OPEC threats. Immediate threats of reduced supplies reinforced the long-lingering fears prompted by blockades during the two World Wars and by OPEC and Anglo-Saxon dominance of the industry. Blockades and OPEC represented disruption of physical supplies; later OPEC action represented, in the eyes of most European governments, oligopolistic pricing practices.

On the other hand, there is evidence that the oil MNCs could deliver cheaper oil and minimize supply disruptions more easily than could the NOCs. In France, for instance, CFP, in 1975, imported its crude at $3.50 more per ton than did one of the top three largest oil MNCs. Another large oil MNC, Esso (France), also in 1975, 'paid FF380/ton [for crude] compared with the national average of FF388.'[21] In 1975, the French NOCs controlled 52 per cent of the market. Another source confirmed these price discrepancies by noting that CFP and ENI – and consumer governments themselves – 'have made deals less favorable from the standpoint of price than those of the U.S. [oil MNC] firms.'[22] Moreover, there is little reason to doubt that this situation will change in the foreseeable future. The OPEC countries – their rhetoric notwithstanding – wish to sell their oil on the most advantageous terms. Therefore, there is no reason that they should favor NOCs. Most of the OPEC countries have huge quantities of crude oil to market and prefer to market consistently. Hence they would be expected to turn to buyers who would take these quantities over a long period of time. The OPEC countries are also used to dealing with the oil MNCs. OPEC countries need the downstream operations that are dominated by the oil MNCs and through which the bulk of the oil has to move. OPEC countries may make occasional exceptions for express political purposes, but these are likely to be minor and infrequent.

The oil MNCs are, with the exception of Royal Dutch/Shell and Mobil, long on crude oil.[23] While the oil MNCs no longer obtain oil under terms of the now-superseded concession system, they still have better access to inexpensive crude than do the NOCs. BNOC and STATOIL are the only NOCs long on crude; in fact, they were established to handle the crude found in their respective portions of the North Sea. CFP and, with its preferential supply agreements, SNEA are almost self-sufficient in crude. However, ENI and Veba are chronically short of crude and have to cover their deficits with arm's length purchases. By the early eighties, both BNOC and STATOIL have access to North Sea crude, which is secure but expensive. Moreover, and very importantly, the oil MNCs are still all larger than the NOCs and have a much more diversified crude supply base.

The oil MNCs also realize a gross profit on their crude that is not available to the generally crude-short NOCs. The new relationships between producing countries and oil MNCs include 'buy-back' prices, service fees, production payments of various designations, and side agreements that tie remuneration to the cost of oil to the companies. The crude profit is much smaller in the late seventies than it was during the sixties. Nevertheless, profits are still available fully to the oil MNCs and only partially to the NOCs. Thus, in all likelihood, the oil MNCs will be able to serve their markets at least as securely and as inexpensively as the NOCs.

We may contrast the European (that is, French, German, and Italian) approach to securing crude oil supplies via NOCs with the Japanese approach.[24] Japan does not have a national oil company. Through the Japan Petroleum Development Corporation (JPDC) established in 1967, the state participates indirectly in about fifty relatively small private companies that are exploring for oil. In all cases, JPDC's equity is less than 50 per cent and in no case does JPDC exercise control. The oil MNCs active in Japan (all foreign, although they frequently act with private Japanese partners) are not unregulated. Under the Oil Industry Law of 1962, importing, refining, and marketing are subject to permission granted by the Ministry of Trade and Industry (MITI). Also, the oil companies' production and import programs have to be reported to MITI. Such programs are either approved or modified in accordance with the oil-supply program prepared by MITI. Armed with these indirect yet very effective controls, Japan seems to be able to obtain supplies as securely and as inexpensively as have the European countries with their NOCs. No European country has, however, anything resembling the omnipresent MITI or the very close government–industry cooperation that exist in Japan.

Assuring Markets

The argument that NOCs are essential because of the oil MNCs' 'market selectivity' is paradoxical. Prior to 1971, the oil MNCs realized large production profits, so they supplied all markets regardless of downstream profitability; low profits or even losses in refining and marketing were more than compensated for by profits realized upstream.[25] Since 1971 the oil MNCs' upstream profits have been drastically reduced, and they have therefore become more concerned with the profitability of their downstream operations. As a consequence, the oil MNCs have abandoned certain markets where they incurred a string of sizeable losses and saw no financial improvement in sight – mostly in countries whose governments kept prices too low.

A case in point is Italy. As long as British Petroleum, Shell, and Esso registered only moderate losses in Italy, these losses were covered by the parent companies from upstream profits. Such compensation effectively transferred funds into Italy. But these companies eventually judged that their losses were intolerable. BP pulled out in 1973, Shell followed in January 1974, and Esso threatened in November 1976 to sell its operations. With BP's and Shell's departure fund transfers ceased. Therefore, BP's losses have been transferred to Monti (BP's buyer), a private Italian oilman; Shell's losses have become those of ENI (Shell's buyer), the Italian NOC. Shell's former losses now have to be covered by the Italian treasury either through higher taxes or through foregone expenditures elsewhere. Italy's situation indicates that while large importing countries must not become victims of MNC selectivity,[26] NOCs might be better able to support their markets if state energy policies did not encourage the oil MNCs to abandon these markets.

Securing Investment Capital

Equally paradoxical is the argument that NOCs are necessary because MNCs are unable or unwilling to invest the huge sums needed for future energy expansion. It is certainly true that since 1971, because of reduced production profits, MNCs have become more selective in their investment policies. Yet the MNCs still obtain more upstream profits – because of their access to low-cost crude – than do the NOCs. Also, their worldwide marketing networks enable MNCs to dispose of larger quantities of crude than can NOCs. The funds available to MNCs for future investments are therefore bound to be considerable. It is likely that MNCs will be willing to commit these funds, although they will continue to draw back from politically high-risk ventures. NOCs claiming to have primary political responsibility for oil supplies may wish to undertake such high-risk investments with adequate government support. That such ventures, though likely to be necessary, will be costly for both the NOCs and their governments is apparent.

Defining 'political risk' in this context is difficult. I refer not to countries where chances of expropriation are great, but rather to countries where ceiling prices are kept 'artificially' low or tax structures are changed retroactively. Also, in investment evaluation, a great deal depends on which phase of the industry's integrated activity the investment is being considered. The argument that NOCs are necessary because governments guarantee their investments has greater merit regarding exploration ventures in high-political-risk areas than concerning ventures in downstream operations.

As a matter of fact, downstream operations in Europe suffer from excess refining and marketing capacity. Refining profit margins are inadequate. In July 1976 five European companies (CFP, ENI, Veba, SNEA, and Petrofina) addressed a joint memorandum to the Commission of the European Communities seeking its approval for their coordinating exploration, production, refining, and marketing policies.[27] The memorandum emphasized the need to coordinate refining expansions and to establish common pricing policies and anti-dumping rules. The memorandum can be viewed as an attempt to improve the companies' overall profitability, with Common Market support. So far, no action has been taken on the memorandum, and none is likely.

Operations of National Oil Companies

Having examined the key objectives of national oil companies, I can now turn to their operating strategies in relation both to the international oil industry and to their home governments. The NOCs' influence on the international oil industry structure has been mixed. After World War II, the defeated Germans adapted to the world oil structure as they saw it, while the French tried to modify that structure to gain greater shares for their companies, CFP and SNEA. Neither the French nor the Italians were successful in changing the world oil pattern,[28] although they were successful in changing the patterns in their own countries. In crisis situations, such as that of 1973–74, even the latter clause is not entirely correct. For instance, CFP and SNEA were no more able than French affiliates of the oil MNCs to respond to the government's urging that they give their home market priority over the foreign markets they also served.[29]

Attitudes of MNCs to NOCs are also mixed. In France, for instance, the NOCs control more than one-half of the market. The government's energy policies are marked by clarity of purpose, and the MNCs are satisfied because they themselves were profitable. In Italy, on the other hand, the NOC controls about 40 per cent of the market. Government policy has been inconsistent; thus some MNCs have left and others would like to withdraw.

Two key elements determine the NOC's relationship with its government: (1) the NOC's actual internal behavior and (2) the behavior of the government toward the NOC. The NOC's internal functioning can be examined by focusing on the way that internal decisions have been made and on the personnel who have made them. One can look at the specific effects that decisions have on profit, pricing, market share, and capacity utilization. Furthermore, internal decision making can be evaluated in terms of the pricing of capital for accounting purposes, the composition, level of, and sources of crude oil and gas, the company's geographic dispersion, and its product diversification.

The authority of governments over NOCs has included taxes, subsidies, the formal charter under which the NOC was established, government membership on the boards of directors, and the state's power to appoint or remove personnel. The governments have also influenced NOCs indirectly by manipulating political pressure and public opinion and by imposing controls that operate indiscriminately throughout the economy. Governments have required that NOCs operate in accordance with specific socioeconomic and political policies. For example, the government may specify levels and costs of crude supplies. It may require financial self-sufficiency or encourage new technologies and products. It could force the company to assist in the attempt to equalize incomes geographically or to act as an instrument in foreign economic ventures.

If the objectives of the NOCs are not quite clear either to the NOC's management or to the government, the relationship between government and NOC becomes ambiguous and counterproductive. Moreover, if the NOC has multiple objectives, as most do, the question arises of tradeoffs among objectives. If ROI is not the overriding consideration, and if the government judges that other aims contribute to the state's economic welfare, who determines the tradeoff? The NOC? The government? Do they decide jointly, or are decisions made by default? Obviously, the answers vary. The basic issue we are concerned with here is whether the method of ownership or control influences the extent and the way in which the services are performed.

The conflict may well be illustrated by the performance of the government-owned British Gas Council (BGC) and the (British) National Coal Board (NCB). BGC determined how much North Sea gas it needed and set an attractive price on that quantity. Gas in excess of that volume was quoted at such a low price that more gas was not produced. However, it might have been in the government's interest to have the gas produced and the surplus exported to the Continent. This was not done. In the early seventies, the BGC began to quote lower prices to gain a share in the industrial market at the expense of coal. Consistent urgings from the NCB brought the government to require BCG to keep its prices up, which would slow the rate of increase in production.

The state's choice of personnel has sometimes exacerbated this conflict between public and private objectives. To create dynamic and competitive NOCs, many governments chose venturesome entrepreneurs for helmsmen. This criterion was used for five of the six NOCs under review in this book. Victor de Metz of CFP, Pierre Guillaumat of SNEA, Enrico Mattei and Eugenio Cefis of ENI, Lords Kearton and Balogh of BNOC, and Arve Johnsen of STATOIL represented a new and specially gifted breed of chief executive officers who feel equally at ease in the corporate boardrooms, in the halls of governments, and some even in the public arena. These executives have wished to create commercially successful enterprises that are as free of government control as possible – a desire that, of course, made the executives less keen to deliver the noncommercial services the governments demanded. Thus, these leaders tended partially to defeat the very purposes for which they were chosen.

We can see from these two examples that the issue of how much control governments should exercise over their NOCs offers an inherent dilemma. 'Excessive' government controls may produce NOCs that are merely extensions of the civil services rather than independent firms with entrepreneurial flair. 'Insufficient' controls may allow the NOCs to become more like private oil companies, which are not interested in delivering noncommercial objectives. The more horizontally and vertically integrated and the more multinational the NOCs become – all patterns characteristic of the oil industry – the less control their governments may have.

The growth of national oil companies has affected their relationships to their governments. All the NOCs started out, obviously, by seeking and obtaining government support. More often than not, this support was explicit. Country to country variations, however, are important. In France as well as Italy government presence in big business is ubiquitous – even if big business is privately owned, or is under mixed (private–public) control. Moreover, in France, governmental influence is bolstered by informal elite links among the *anciens* of the *grandes écoles*, there being a regular migration of officials from government to business and vice versa.

However, in the five countries under review, and even in France, managers of NOCs have attempted to develop their own power and independence, to control the market or create new markets both at home and abroad, occasionally to maximize profits, and, almost invariably, to grow. Emphasis on growth comes from two sources. Within the company, executives seek the higher pay, greater promotion opportunities, job security, and, perhaps most important, the prestige that accompany larger organizations. Executives also realize that the larger the firm, the greater its ability to influence or even to control its social and political environment. As NOCs' power in their home countries becomes more concentrated, they may become less eager in serving the public and more interested in serving their own corporate goals. When in 1974 the French government imposed ceiling prices on petroleum products (prices that, in the eyes of the companies, were too low), an executive of a French NOC commented that he wished he could operate as Veba does in Germany – that is, without government interference. Some analysts hypothesize that 'large-scale public enterprise and widespread public regulation may be incompatible.'[31]

One observer summarizes the power play between NOCs and their governments thus: when CFP, ENI, and SNEA 'have the government's wind at their backs, they set national sails, but when the wind blows in their faces they insist on company interest.'[32]

Multinationalization of National Oil Companies

The initial objectives of national oil companies did not include their joining the multinational oil companies' oligopoly. However, as the NOCs grew, geographical diversification became necessary to minimize technical and political risks. If an oil MNC registers losses in, say, Germany, it may offset these losses in the Philippines. NOCs, operating primarily in their home markets, cannot spread their risks. Hoping to offset their losses, some NOCs became multinational. ENI, the NOC which for fifteen years was the real or alleged thorn in the side of the oil MNCs, began to advertise itself as a multinational company in 1969 and has emphasized this aspect ever since. In France, CFP and SNEA tend not to say openly that they are MNCs, but in fact they are. CFP, ENI, SNEA, STATOIL, and BNOC all grew up in protected, hothouse home markets, venturing into international operations from protected home bases. In foreign countries the NOCs offered a differentiated product – differentiated neither by price nor quality but by political context: it was nonprivate, non-Anglo-Saxon, and sometimes nonforeign. As early as 1974, one OPEC official referred to the NOCs as being 'just like other international oil companies.'[33] STATOIL will soon become multinational, because the small size of the Norwegian domestic market will force the company to sell its crude abroad. Veba's and BNOC's future international roles are not yet certain; because of the large size of the British market, BNOC would have a ready domestic market for its crude. However, British North Sea crude is, technically, more suited for markets where it would command a price premium such as the US market, whereas the British market could continue to be well served by Persian Gulf crudes. Thus it is likely that BNOC will soon go multinational. It is already becoming increasingly difficult to talk about NOCs versus MNCs. The 'Seven Sisters' now number ten and may soon be twelve or thirteen.

References and Notes

1. The coining of the phrase 'Seven Sisters' has been attributed to Enrico Mattei, the founding chief executive officer of the Italian National Oil Company, ENI.
2. One book is Paul H. Frankel's *Mattei: Oil and Power Politics*, Praeger, New York, 1966.
3. Standard Oil of Indiana, Atlantic Richfield, Continental Oil, Phillips Petroleum, and Occidental Petroleum were the largest independents.
4. 'International Corporate Scoreboard, 1977', *Business Week*, July 24, 1978, pp. 82–120. For share of European refining capacity see W. Monig, D. Schmitt, H. K. Schneider, and J. Schurmann, *Konzentration und Wettbewerb in der Energiewirtschaft*, R. Oldenbourg Verlag, Munich, 1977, pp. 142–144.
5. Leslie E. Grayson, *Economics of Energy*, The Darwin Press, Princeton, 1975, pp. IX–XIII.
6. Louis Turner, *Oil Companies and the World System*, The Royal Institute of International Affairs, London, 1977.
7. *BP Statistical Review of the World Oil Industry, 1970 and 1977*, BP, London, 1971 and 1978.
8. Paul H. Frankel, *Structural Analysis of the International Oil Industry*, International Bar Association (Business Law), London, 1978.
9. Lawrence G. Franko, *The European Multinationals*, Harper and Row, London, 1976, p. 52.
10. 'Drilling New Ground', *Wall Street Journal*, November 23, 1977.
11. Edward M. Krapels, *Controlling Oil; British Policy and the BNOC*, Publication No. 95–99, US Government Printing Office, Washington, 1977.
12. Public Accounts Committee, *North Sea Oil and Gas*, Session 1972–1973, HM Stationery Office, London, 1973, p. XXIV.
13. Louis Turner, *Oil Companies and the World System*, The Royal Institute of International Affairs, London, 1977.
14. Robert B. Stobaugh, 'The Oil Companies in Crisis', *Daedelus*, Fall 1975, p. 189.
15. *Vision Magazine*, Paris, October 1975. *Vision* classified sales by major industrial sector. If a company was listed as both a petroleum and chemical company, it is listed as such here and so, of course, with the other two categories.

Type of company	No. of companies	Sales	Total (per cent)
Chemicals only	10	61,838	42.7
Chemicals and petroleum	3	57,899	40.0
Petroleum	4	25,074	17.3
		144,811	100.0

16. The literature on this point is voluminous. Academic contributions include W. G. Shepherd, *Economic Performance Under Public Ownership,* Yale University Press, New Haven, 1965; R. E. Caves *et al.*, *Britain's Economic Prospects*, Brookings, Washington, 1968; Ralph Turvey, *Economic Analyses and Public Enterprises*, Rowman, London, 1971; Clair Wilcox and W. G. Shepherd, *Public Policy Towards Business*, Irwin, Homewood, Ill., 1975; W. G. Shepherd, (ed.), *Public Enterprise; Economic Analysis of Theory and Practice*, Lexington Books, Lexington, Mass., 1976; John Sheahen, 'Government Competition and the Performance of the French Automobile Industry', *Journal of Industrial Economics*, June 1960, pp. 197–215; Merrill and Schneider, 'Government Firms in Oligopoly Industries', *Quarterly Journal of Economics*, 1966, pp. 400–412. One of the best popular summaries was published by the *London Times* in a 'Special Report on the Principles, Planning, and Operation of a Mixed Economy,' June 25, 1975.

17. John Vinocus, 'A Triumph in Sweden,' *New York Times*, May 28, 1978.
18. Paul Lewis, 'Jet Lag at Air France,' *New York Times*, September 18, 1977.
19. See *London Times*, *op. cit.*, for the United Kingdom. For Sweden, see Adrian Hamilton, 'Problems of a State Owned Conglomerate' (The Statsforetag Group), *The Financial Times* (London), August 25, 1975. For Norway, see Robert F. Norden, 'The Norwegian State Railways; Profit-Motivated Enterprise or a Community Service?', *Long Range Planning*, April 1978, pp. 13–18.
20. *Business Week*, March 3, 1973, p. 68.
21. Private communication to the author, Spring 1976, and 'France,' *Petroleum Economist*, April 1977, p. 153.
22. Paul H. Frankel, 'Statement,' *Multinational Oil Companies and OPEC*, Hearings, US Senate, June 3, 1976, US Government Printing Office, Washington, 1977, p. 106.
23. Various issues of the *Annual Reports* of the oil MNCs comparing crude production with refinery runs and (adjusted) product sales.
24. The International Bar Association in 1975 sent out a questionnaire on the 'Patterns of State Intervention in the Oil Industry.' The answer for Japan was provided for by Professor J. Tsukamoto of Chou University's Law School. (The questionnaire for all countries was analyzed and summarized by Jean Devaux-Charbonnel and Adrian D. G. Hill, in the *International Business Lawyer*, January 1976.)
25. M. Adelman, *The World Petroleum Market*, Johns Hopkins University Press, Baltimore, 1972.
26. 'The Role of Consumer-Country National Companies in the New Petroleum Industry Situation,' ENI, Presentation to the International Energy Agency, Paris, March 31, 1976.
27. Commission of the European Communities, Brussels, July 23, 1976 (mimeographed) and summarized in C. H. Farnsworth, 'Oil Firms Seen Challenging Seven Sisters,' *International Herald Tribune* (Paris), September 24, 1976.
28. Horst Mendershausen, *Coping with the Oil Crisis*, Johns Hopkins University Press, Baltimore, 1976, pp. 35–37.
29. Horst Mendershausen, *Coping with the Oil Crisis*, Johns Hopkins University Press, Baltimore, 1976, pp. 63–65.
30. F. Ghadar, *A Study of the Evolution of Strategy in Petroleum Exporting Nations*, unpublished doctoral thesis, Harvard Business School, 1976.
31. James Q. Wilson and Patricia Rachal, 'Can the Government Regulate Itself?', *The Public Interest*, Winter 1977, p. 14.
32. Mendershausen, p. 36, and similar sentiments repeatedly expressed to the author, 1975–77.
33. Lawrence G. Franko, *The European Multinationals*, Harper and Row, London, 1976, p. 65.

CHAPTER 2

The French Oil Industry and Its Regulation

France is unusual in that it has two oil companies in which the government has substantial equity interest. Both companies, Compagnie Française des Pétroles and Société National Elf-Aquitaine, between them controlling half the French market, are operators on a world scale and are the result of the development over many years of a coherent national oil policy. The primary objectives of this policy have been to ensure security of supply of a vital raw material by providing a protected home market, and thereby giving the national companies the ability to operate successfully. At the same time the stimulus and diversity of competition with private companies was promoted and the national companies were encouraged to operate internationally on a scale sufficient to make them a significant presence.

That this has been achieved is itself the mark of success of the policy, although in 1980 the immediate future was less certain following the upheaval of the energy crisis, the new energy requirements of France which resulted, and various moves at the level of the Common Market designed to discourage the kind of direct assistance which the government had given to its national companies.

Early History

From the beginning of the 20th century until World War I the French oil market grew steadily and unremarkably, the country depending entirely on imports of all products, crude and finished.

Oil consumption[1]	
1900	320,000 tons
1913	700,000 tons

Moves to establish a domestic industry met with no success, as did efforts to secure supplies, and short-sighted fiscal policies effectively reduced French refineries to 'ticking over' operation only.[2] World War I and the consequent disruption was such that in early 1918 it looked as if France would, in March, literally run out of oil, and only prompt action by President Woodrow Wilson of the United States reacting to a telegram from Clemenceau averted this situation. In consequence the government set up, in August 1918, the Commissariat Général aux Essences et aux Combustibles to develop a petroleum policy in response to the awareness of the importance of oil to the nation. The first result from this body, or from those who negotiated in its name, was an amendment in January 1921 to the San Remo Treaty of the year before, ceding to the French state the share of the Turkish Petroleum Company (TPC) held until then by the Deutsche Bank (25 per cent). The company was at that time controlled by Anglo-Persian (later BP) and on the invitation of the British in 1923, the French government set about constituting a commercial group to control the shares and take part in exploration activities, which had not yet started. In view of the private nature of the other partners in TPC and the national interest involved, the government decided to associate public interest and private capital and asked an industrialist, Ernest Mercier, to bring together the necessary private investors 'to create an instrument suitable for carrying out a national oil policy'.[3] By March 1924, Mercier had brought together sufficient investors to launch the company to take up the TPC shares despite their suspicion of the state's interest. It was called Compagnie Française des Pétroles. In the following year, 1925, the government set up an administrative and policy organization, Office Nationale des Combustibles Liquides (ONCL), to encourage exploration by forming and investing in exploration companies with other partners, usually private. This was to take advantage of a 1922 law giving applicants exclusive exploration rights on French territory, both metropolitan and overseas. While this activity was taking place various changes were occurring in TPC as moves were made to admit two American companies, Standard Oil of New Jersey and Standard Oil of New York, and reward Gulbenkian for negotiation on behalf of the company. The end result was a small reduction in the CFP shareholding to 23.75 per cent in 1927, and the company renamed itself Iraq Petroleum Company (IPC) and struck oil at Mosul, Iraq. CFP now had oil to sell in France.

The French market since World War I had been the scene of fierce competition, mostly particularly between Shell and Esso, who between them controlled nearly all supplies (Shell 15 per cent, Esso 56 per cent) and two-thirds of the market. French marketers were hard pressed to stay in business, and strong political feeling against the outside companies, together with continuing development of policy, led to the choice being formed in the Assemblée between monopoly or import control with quotas[4] for regulation of the industry. In 1928, of two laws passed in March, the second option was taken.

The 1928 law set out to achieve two purposes, the security of supplies and the development of a national oil industry, particularly refining. This was to be achieved without the direct participation of the state except by import control, and controls were of two sorts: those covering finished products, essentially protecting marketers, and those covering crude imports, protecting refiners. The crude import quotas (A20) were awarded in 1931 for twenty years to eleven companies who had all to undertake obligations in order to qualify. These were the maintenance of compulsory 'reserve stocks', the acceptance on request from the government of 'contracts of national interest' for crudes, products, etc., in the import quota, the construction of the necessary equipment, French management, French flag ships for 66 per cent of the imports, and the manufacture of products required by the Army. The French groups received 54.2 per cent of the quotas.

The quotas for finished products (A3) were awarded to a larger number of companies (fifty-one) and were valid for three years for limited quantities of petrol and lubricants, unlimited for other products. The total amount of authorizations was larger than the market (which by 1929 amounted to 3 million tons of crude).[5] An important principle governing the (A3) quotas was that companies were not allowed to be net importers of finished products, providing a further incentive for French refining. The laws became the government's principal control over the market and policy developed much further in later years. The *modus operandi* established in 1928 was to shape the French industry for years to come.

An early outcome of the laws was a change in CFP's statutes. There was considerable opposition from the shareholders of CFP, many of whom were distribution companies, to the idea of establishing a refining subsidiary to absorb the oil from Iraq and take advantage of the new law. This led the government to take a shareholding, both in CFP (35 per cent shares, 40 per cent voting) and in the proposed refining subsidiary, Compagnie Française de Raffinage (CFR) (10 per cent shares). The change, which was first proposed in 1928, aroused considerable political opposition and was not ratified until July 1931.[6] When finally agreed, it confirmed the government's overall control of CFP and gave CFR the right to refine 25 per cent of French consumption (there is a full account in the case of Compagnie Française des Pétroles). CFR thus became CFP's outlet in France and it is noteworthy that the theoretical relationship of CFR to CFP had much in common with the relationship of the foreign multinationals to their French affiliates. This enabled the government to maintain an even-handed control of the French market at a distance by use of the import quotas without directly marking out CFR for special treatment.

Throughout the thirties CFP and CFR were engaged in building up their organizations. Following the 1928 law, fifteen refineries were set up in France with two of the largest belonging to CFP.[7] Development continued apace and some notable technical advances were made by Frenchmen. In 1922, Conrad Schlumberger had perfected a method of electric prospection of wells and by 1934 the method was being successfully used by Société de Prospection Géophysique, with which CFP was associated, and which was later to become almost unrivaled worldwide in its field. In 1937 a major improvement was made to refining with the invention of catalytic cracking by another Frenchman, Houdry.[8] ONCL, in combination with some of the colonial administrations, encouraged and formed various exploration teams but no concrete result from these came until 1939 when Centres de Recherches de Pétrole du Midi (CRPM) struck gas in Southwest France at Saint Marcet. CFP, however, in both developing its existing finds and extending exploration, was achieving success in the Middle East which was rising fast in importance as a source of oil. In the twenties only 17 per cent of France's supply came from this area; by 1938, after a sixfold increase in consumption, the figure was 49 per cent.[9] To carry the oil and augment CFP's efforts in building up an oil fleet, ONCL had invested in 30 per cent of Société Française de Transports

Pétroliers (SFTP) in 1938, but the outbreak of World War II brought activities to a halt.

During the war not much was achieved. The find at Saint Marcet had prompted a new initiative from the government when it created Régie Autonome de Pétrole (RAP) to exploit the gas and invest in further activity. RAP, set up in 1939 and ratified in 1941, was innovatory, in that it was the first government-financed body which had a direct industrial and commercial role; bodies such as ONCL and SFTP could only participate indirectly. One of RAP's functions was to explore further in Aquitaine but neither it nor a new company set up by the Vichy government for the same purpose, Société Nationale des Pétroles d'Aquitaine (SNPA), made any discovery during the war. The constitution of SNPA was in keeping with a principle of the national oil policy of uniting private and public interests. The state held 51 per cent of the shares and private investors held 49 per cent.

The Post-War Reorganization and Expansion

World War II again served to underline the importance of oil to a modern nation and for General de Gaulle, newly in power, the importance of independence of supply. If the pre-war policy aim could be abbreviated to the importance of establishing a French refining industry, the post-war aim was to acquire French oil sources. In reality the policy had sought and would seek to strengthen the French influence in all sectors of the industry, but the accent was certainly different. It was the government's judgment 'that the fact of buying from others, or the need to buy from others, regardless of the terms at which one buys, is itself dependence'.[10] Accordingly two important innovations were made. To encourage research, both pure and applied, training, and information gathering, the Institut Français de Pétrole (IFP) was set up in 1943; it was to be financed in its work by a sales tax on petroleum products. Then in 1945 the state created an 'établissement public', or public company, Bureau de Recherches de Pétrole (BRP), wholly state owned, and with a wider brief to prepare and put into practice the government's policy with regard to exploration, infrastructure development, and the creation, in conjunction with private groups, of oil industry companies with majority French holdings to operate in French-controlled territory. (For full details see Chapter 4 on Société Nationale Elf-Aquitaine.) Like ONCL, BRP was to intervene and to act as overall controller of policy. Shares of most, but not all, the companies in which the state already had an interest were made over to BRP. The decree instituting BRP said: 'In time of peace as in war oil is an indispensable primary product in the economy of an important nation.... Oil produced and refined in France (this meant French territories) should be the objective.... This solution is the only perfect one due notably to its considerable economic advantages.'[11] Shortly preceding this decree, the state had also ('Ordonnance 45–1483') given itself the power to fix prices of products and services in the petroleum field, and hence, with the import controls established in 1928 (and then in abeyance due to the disruption of the war), the state now had all powers in hydrocarbon matters. In the following decade there developed an intensive search for French oil all over the French Empire. French oil (and by French one must understand the French territories as well) was not discovered until 1949 when SNPA found oil in Aquitaine, although the size of the field (3.5 million tons) was far smaller than the size of the French market (12.3 million tons including exports of 1.5 million tons). In 1951 SNPA found another gas field in Aquitaine and was now, with RAP, in a position to supply a substantial part of the French market. RAP was already distributing in Southwest France and its network had been specifically exempted from nationalization as belonging to a company which contained a state interest of 50 per cent or more. SNPA too was allowed to develop a network serving Central and Northern France as far as Paris, which was to handle a volume of gas many times that produced by RAP.

In 1950 the government began to tighten its control of the oil industry and the import quota regime was reimposed. The administration was altered from the situation created by the 1928 law with the crude import licenses (A20). Slight variations had previously occurred in response to newcomers, mergers, successful marketing efforts, and so on, but the new licenses were valid for thirteen years (A13) and the amount awarded to French groups dropped slightly to 49.6 per cent (from 54.2 per cent). The control of government was nonetheless tight and indeed the period since 1928 had seen a steady increase in control and influence on the domestic market with renewal of each authorization of import quotas for finished products (A3). On May 27, 1950, the government also introduced a new petroleum products sales tax ('Fonds de Soutien'), for which the original intention was that it would subsidize synthetic petrol production and national oil production generally. Control tightened further in 1952 when an 'ordonnance' was passed obliging refiner–distributors to refine 90 per cent of their sales in France, thus leaving only those independent distributors free to take real advantage of the A3.

The end of 1953 brought a change in the use of the 'Fonds de Soutien' when a large part of the proceeds of the tax was earmarked for BRP, although money also went to subsidize home production and to the general budget. BRP up to this time had existed on grants to make its investments in exploration companies which were now working in Equatorial Africa, Algeria, and France. Investments had often been made in conjunction with private investors, but in 1953, with only the two discoveries at Lacq to show for the efforts expended, enthusiasm for this investment was dwindling. In 1954 interest was reawakened when a company controlled by Esso struck oil in Les Landes, France. Although subsequent production was smaller (0.8 million tons per annum), initially the find was thought to be a large producer (3 million tons per annum) and it was a big stimulus to investment, which nonetheless was taking place more outside Metropolitan France than in it. In Equatorial Africa BRP had formed companies in conjunction with Mobil and Shell. In Algeria BRP was working with CFP, Shell, and a range of American and European 'independents'. The policy adopted was essentially 'open door' and all companies operating in France were specifically encouraged to participate, but only Shell, which did not have large Middle East resources, showed any real interest. CFP took an active role in Algeria, but this was nonetheless secondary to its Middle East interests and its preoccupation with finding markets for its fast growing oil supplies. The exploration efforts began to bear fruit in profusion from 1956 when strikes of oil were made in Gabon and of oil and gas in Algeria, and a year later of oil again in the Congo. The effect of these finds on French thinking was euphoric and commentators spoke in terms of a French Middle East. The reality proved to be less impressive, but nonetheless the finds were to have an important effect on the organization of the French oil industry.

The Emergence of Franc Zone Crude

Protected by the quotas, the oil industry had built up prosperous refining and marketing operations in France, and for refining had instituted a certain amount of 'industry agreements'. These covered three products, petrol, gas oil, and fuel oil, and were derived directly from the authorization (A3). The agreements for gas oil and fuel oil were based on the market position and authorization for petrol. Refining quotas differed from market positions, as CFP, while having the right to 25 per cent of refining, did not have a product market position of the same size.[12]

Notwithstanding this comfortable arrangement, the independent distributors were not prospering in the face of competition from the foreign multinationals and in 1955 some of them regrouped under the trademark TOTAL. This was CFP's trademark and its adoption by the independent retailers, in many of which CFP had shareholdings, gave CFP, via CFR, its first direct outlet to the French products market.

The major oil strikes overseas brought with them a change in government policy, directing the marketers to buy more and more Franc Zone oil. In 1957 for the first time the authorizations (A3) were not larger than market requirements, and this plus the requirements of every importer to reveal to the Ministry its source of imports enabled the Ministry to orient imports toward the new sources. The Ministry was in a position to refuse foreign exchange for an import it considered unsuitable. It was not in fact until 1959 that the first ton of Algerian oil reached the market, and not until 1961 that Franc Zone oil was removed from import control altogether, by which time various small finds of oil had also been made in France, principally in the Paris region. In 1960 the brief of BRP was widened to include encouragement of production companies and a refining company, Union Générale des Pétroles (UGP), was set up with BRP as the controlling shareholder through a grouping of its companies called 'Groupement des Exploitants Pétroliers', together with RAP (which it also controlled) and SNRepal (a joint company with the Algerian government which held 51 per cent of SNRepal's shares). UGP took a 60 per cent share in the refinery at Ambes (Bordeaux) belonging to Caltex (owned by Standard Oil of California and Texaco), the joint company calling itself Union Industrielle de Pétrole (UIP). UGP also took over completely Caltex's French distribution, thereby gaining an immediate 4 per cent market share. The justification for UGP was that it provided the French production companies with an outlet for their oil, which in a buyers' market was already expensive, prices for a landed metric ton being of the order of FF96.6 and for a landed metric ton from Kuwait being FF82.0 (1963 prices).[13]

However, under the 1928 laws the government already had powers ('devoir national') to oblige the refiners to buy from a specific source and full exercise of this authority would have made the creation of UGP unnecessary. Nonetheless, despite UGP, 'devoir national' obligations were imposed on all French refiners to buy 70–80 per cent of all Franc Zone crude, obligations that were reduced by the amount that any company (principally Shell) was already producing on French territory. This imposition aroused considerable opposition from the foreign multinationals, and to a certain extent from CFP as well, and the reason was the price of oil thus sold. There was a strong feeling that the price of Franc Zone oil was artificially high and consequently purchase of it effected a subsidy from the other marketers to the French companies. As both Shell and CFP had production in the Franc Zone, they would have been aware of true production costs. The price of the oil was fixed as half-way between the price of open market marginal crude sales and the transfer price of producing companies selling oil to their French affiliates, and it was in consequence an astute squeeze on company profits.

Volumes produced in the Franc Zone grew rapidly and, by 1964, 41 per cent of French imports came from Africa, whereas, in 1955, 94 per cent had come from the Middle East[14] (see Appendix 1). In 1963 the crude import licenses again came up for renewal (for application in September 1965) and this time the period was shortened to ten years. Now for the first time the quotas were less than the market, in line with the decision two years earlier to free Franc Zone crude from import control. For the affiliates of international companies the amount authorized was equivalent to their 1957 import tonnage, though during the intervening period imports of crude had grown at 12 per cent per annum average. Nine companies were given quotas. One, UGP, was new; the others were Esso, Mobil, BP, Shell, Compagnie de Produits Chimiques et Raffineries de Berre (PCRB), UIP, CFR, and ANTAR, a French independent. In all, French companies now had 61.3 per cent of the quotas, the rules governing which had been tightened allowing for a five-year review, maximum 15 per cent increase, and submission of supply programs to DICA (Direction des Carburants). It is of note that between 1950 and 1963 the successive increases in the permitted import tonnage had been 20 per cent, 10.25 per cent, 7.5 per cent, and 5 per cent of the amount of finished products allowed (A3). A secondary objective of UGP was to introduce more competition into the refining and distribution market, and in 1964 the industry agreements broke up following an initiative by Esso. This was prompted by the behavior of UGP, which had not been party to the agreements and was using the situation to increase its market share. The agreements had related, by means of quotas based on the import authorization, the market positions and refining capacities of three main products: petrol, gas oil, and fuel oil (see page 43). This brought a strong incentive to limiting competition and meant that any increases of position had to be brought through mergers, not achieved through competition.

The boom of the sixties proved an excellent vehicle for the development of

French oil policy. With a supply of French oil, control over the home market, and two strong groups, CFP and BRP, expansion took place steadily. Political uncertainty in North Africa, normal commercial prudence, and the continued search for security of supply kept the pressure on exploration, which went further and further afield and outside the Franc Zone. IFP, which was benefiting from ever larger sums from the sales tax which funded it, began to spread its activities beyond pure documenting, training, and research in France. In 1959 IFP set up an engineering firm, TECHNIP, to handle plant design and other studies, and it began to take on training and technical missions overseas which were part of a policy aim of increasing French influence in petroleum-producing areas. It was also devoting considerable amounts of effort to exploration, production, and refining research. Its investments in offshore technology proved in later years to be particularly worthwhile. Of course as a service to the industry as a whole in France, IFP's output was available to everyone, but by virtue of the foreign multinationals frequently choosing to rely on their research facilities, the national groups benefited disproportionally from the work of IFP. An alternative view was that the MNCs chose to rely on their own research because it was better than that of IFP and, hence, the national groups suffered by comparison.

Exploration was not the only activity that diversified geographically. Refineries under French control were set up in other countries in the Franc Zone (Senegal, Ivory Coast, Gabon, Madagascar, Martinique, and Cambodia) and French companies specializing in drilling, geophysics, and equipment became well known across the world. For the state-controlled section of the industry a significant change took place in 1965 when BRP and RAP were fused into one organization, Etablissement de Recherche et d'Activités Pétrolières (ERAP). This was effectively both a holding and an operating company of all the activities which had previously been controlled by BRP. In the reorganization that followed, the Elf trademark appeared in April 1967 to cover activities of the new group and UGP became (in 1967) Elf Union. The Elf trademark replaced seven different marks.[15] ERAP essentially had two sides to its activities. One was to control the government's holding (51 per cent) in SNPA (whose shares had previously been held by BRP), which was now a major gas and sulfur (from the gas) producer and an important supplier of, and explorer for, crude oil. Second, ERAP controlled a refinery and distribution network represented previously by UGP and UIP, together with the exploration and production groups of BRP and RAP. Nonetheless the management of ERAP and SNPA stayed distinct due to the large private shareholdings in the latter.

Judicious use of the A3 authorizations continued to foster the growth of French government or quasi-government companies at the expense of both independent and multinational operators, and the share of the latter of the market had fallen from 80.5 per cent in 1945 to 60.4 per cent in 1955, to 53.4 per cent in 1965. A long, unspoken objective of the government was given formal exposure by the Minister of Industry, M. Bettencourt, in the National Assembly in November 1968. He spoke of three principal policy objectives, namely '. . . to assure the country's control of diversified production meeting its needs, to see the coexistence in France of oil companies while maintaining a stimulating competitive climate, with 50 to 60 per cent of the national market controlled by French companies, and to provide the country with a means of foreign distribution and refining and to assure a favorable commercial balance in petroleum products'.

However, the government's policy was already running foul of the Common Market Commission and the Treaty of Rome with its clauses covering the free movement of industrial goods and its purposes in removing discrimination in favor of commercial national monopolies (article 37, paragraph 1). In 1959 the government had informed the Commission that it recognized the industry as falling under article 37 (despite previous attempts to show that the monopoly was delegated to the companies and they were not covered by the article), and through the sixties various concessions were made to the liberalizing of movements of finished products within the EEC. The moves were essentially cosmetic, and the Commission was not able to press the issue, recognizing it to be a sensitive national policy.

Although the foreign multinationals, in seeing their hold on the French market drop in share if not in volume, had reason to feel aggrieved, the companies that had retreated most were the independent distributors. Many had been originally investors in CFR and, until the difficulties of the early fifties, they had prevented CFR from having any direct marketing outlet. The appearance of the TOTAL brand mark, in 1953, heralded an era when many independent distributors gave up the struggle to compete with the large foreign companies and took refuge under the TOTAL brand by being taken over by CFR. By 1960 only two significant independents were left, Desmaris Frères, which was taken over by CFR in 1966, and ANTAR, which in 1970, prompted by either an inspired reading of the future of the oil products markets or by a gloomy view of French oil policy, was taken over by Elf–ERAP. Both companies had previously held some 10 per cent of the petroleum market and they had been the biggest of the independents, ANTAR being the bigger of the two with fleet, refinery, and distribution investments. ANTAR sold out to several buyers; Elf–ERAP took the controlling share of 41 per cent, but the state took 10 per cent directly, Caltex took 20 per cent, and CFP took 24 per cent. Five per cent was retained by the original shareholder, Pechelbronn.

This was not the first time that the two French groups had come together for investment purposes. They had formed joint investment groups for exploring the North Sea, CFP acting as operator in the British Sector, RAP acting as operator in the Dutch Sector, and BRP acting as operator in the Norwegian and Dutch Sectors. They had also come together downstream in setting up in 1969 a joint company for organic chemicals, Aquitaine TOTAL Organico (ATO), and in 1970 they set up a joint chemical company, Compagnie de Pétrochemie. This was in response to the desire of the government to strengthen the French petrochemical industry.

The end of the decade of the sixties thus saw both French groups in good positions. They held about 23 per cent each of the French market, acquired by CFP by merger and by Elf–ERAP by market expansion and merger, aided by government policy and at the expense of independents and foreign groups. CFP had an extensive foreign network of sales and both companies were accomplished explorers although their resources were spread very wide. IFP, in conjunction with the groups, was building up a network of service contracts and considerable off-shore expertise. The French refining industry was strong and closely related to the company's requirements and, with reference to the past alone, the policy aim of a strong French oil industry seemed most successful. Auguries for the future were less rosy.

The Upsets of the Seventies

The war and subsequent independence in 1962 of Algeria, from where France imported 27 per cent of its oil in 1970, had long been a reason for exploration activities to be directed elsewhere, and nationalization of Shell's assets in Algeria in 1970 was an unfortunate omen for 1971 when all the French groups' assets were nationalized. By virtue of its dependence on Algerian sources for 80 per cent of its supply, Elf–ERAP was more affected than CFP, which only took 20 per cent of its oil from Algeria, and both companies had been beneficiaries of a discount for delivery to France. The immediate aftermath of the event was acrimonious for all parties concerned, and the upset to the French oil industry was sufficient for the Cabinet to make a public declaration of the future development of policy in July 1971. The country was still to have two oil groups, but Elf ERAP would integrate its independent offshoot SNPA and the group would continue to act as a vehicle for special relationships with producer countries, as had been the case with Algeria from the setting up of ERAP – the practice had subsequently been extended to agreements with Iraq and Iran. In 1973 the French oil industry, and the government policy since 1928 of Monopole Délégué, came under strong attack when it was alleged that the companies had been collaborating on prices. A report by Deputy, M. Julian Schvartz, following the oil crisis of that year criticized a number of other practices and, as a result, a protocol was drawn up between the public authorities and the companies to define permissible competition practices.[16] Following on from the criticisms of this period, the Common Market Commission renewed its interest after a complaint to the Commission of unfair treatment by a small French independent, although nothing specific resulted. Changes were also made to various depletion allowances and foreign exchange provisions and the cost to the industry of this alone was expected to be FF 1.3 billion per year.[17]

The oil crisis provided its own measure of disruption to the industry. CFP, strongly dependent on the Middle East, found itself confronted by prior obligations to supply France rather than its other outlets, and the realization of this by some customers was to lead them to be wary of undertaking long-term contracts with the company.[18] Elf, with its resources spread more widely than CFP, rode the crisis a little more easily. But all the firms in France were subjected to price controls on products, controls which did not allow direct translation of the new crude prices to the market. The price controls obviously had a dampening effect on profits, and was as serious as the drop in demand that became apparent in the years that followed, which led to serious undercapacity utilization of refineries.

By 1975 utilization had dropped to 75 per cent or less for both groups, and in an effort to stave off a price war and to draw attention to their plight, the two companies joined with ENI of Italy, Veba of Germany, and Petrofina of Belgium in writing a memorandum to the Common Market Commission in July 1976. The problem was not only refining. Exploration and production, in which both groups were active participants, had become much more expensive as activity switched away from politically inhospitable areas to areas of greater political security, but more often than not considerably greater natural difficulty, such as the North Sea. Here the French groups had been both omnipresent and successful, but the cost of development of their finds was straining their balance sheets when they were making no compensatory profits in their principal market, France. This was despite continued discriminatory action by the government in their favor. In March 1974, a 'tax parafiscale' of FF3.9/hl on petrol (reduced to FF3.00/hl in January 1975) was used to provision the budget of La Caisse Nationale de l'Energie who subsequently recycled it to Elf. The tax penalized those companies strong in petrol (Shell, Esso) against those strong in fuel oils (BP, Elf, CFP), the government's declared reasons for the tax including the provision to France of more expensive Algerian crude (CFP, Elf), the expense of Gabonese production (Elf, Shell), the supply of independents (CFR), and the use of heavier than average barrels (BP and Elf). The 1974 proceeds of the tax were FF690 million. Also an inventory reserve tax, started in the forties had a one-time detrimental effect on the non-French oil companies when they recorded large inventory profits in 1974. As the French groups were not allowed deficits and had made small profits in 1974, they were not taxed. The non-French companies, however, effectively suffered a transfer out of profits and consequently paid more tax.

The French oil policy, however, found itself unable to deal with the full implications of the oil crisis. Continued pressure from the Commission had led the government to reduce the market's protection. But both within it and outside the French groups were competing with the foreign multinationals who still had access to profitable markets, principally the United States, and to cheap crude (or at least cheaper crude). The more that the Elf and TOTAL brand names spread outside France, and half of CFP's sales and nearly a fifth of Elf–ERAP's took

Table 1. 'Tax parafiscale', 1974[19] (millions of francs)

	Payments	Receipts	Net
Elf/ANTAR	168	328	+160
CFR/TOTAL	188	224	+36
BP	59	98	+39
Shell	110	40	−70
Esso	100	0	−100
Mobil	44	0	−44
Fina	21	0	−21

place outside France, the more they would meet strong competition. And yet if they were encouraged to explore in areas like the 'secure' North Sea, profits would have to be generated for investment and only a very high prices and full protected French market could provide this. Politically it was impracticable.

French Oil Policy, Success or Failure?

The year 1976 was marked by the writing of the Brussels memorandum by five of Europe's national oil companies and the final uniting of all the companies in which the government had a stake and that had grown out of or with BRP's activities, into Elf–ERAP and finally Elf-Aquitaine, in which the government stake was 70 per cent.

From the outset the concept of security had been the dominant influence on policy, and while control of the oil industry had always been more extensive than a simple interpretation of the 1928 laws would imply, the voluntary and amicable cooperation of all firms operating in France had been sought. In this the French experience distinguished itself from the Italian experience with their national oil company, ENI, where the policy had been more one of confrontation. By using the system of a delegated monopoly the government had given a clear lead to the industrialists to organize the industry as they saw fit, once the overall framework was established by the government. By controlling the market, the government could ensure not only that everyone had a share of investment expansion, but that such investment gave suitable returns. Until the middle seventies the industry had always been profitable and logistical control by the government had ensured that the costs to France of industrial expansion had been fairly spread around the country.

In terms of its early objectives, refined products in France came as near as possible from French refineries, there was a substantial body of French expertise in all phases of the oil industry, and France possessed, by virtue of the French flag requirements, a modern and efficient oil fleet. In terms of the post-war objectives of the policy, the achievement of secure supply had proved more elusive than might have been expected by General de Gaulle when he took the initiative. But nonetheless Elf had resources in many countries and CFP, while still heavily dependent on the Middle East, at least was not dependent on any one country in this area and was very long on crude supplies as a whole. Elf was about three-quarters self-sufficient in 1976. If supplies were still largely not under direct French control, in the post-1973 atmosphere this was evidently not a weakness. French influence was extensive and while true of production it was more true of marketing. The overseas markets of CFP were the more extensive but in Europe Elf had grown rapidly and between them the groups held about 5 per cent of the Western European market outside France in 1975 (finished product sales: CFP 22.4 million tons, Elf 5.7 million tons).[20] In France Elf and TOTAL between them controlled half the market (in 1975 TOTAL had 27.2 per cent, Elf 14.0 per cent, and ANTAR 9.0 per cent),[21] and it seemed unlikely that the government would allow this hold to break.

By ensuring the profitable presence of the oil multinationals, and their acquiescence in the oil industry administration, beneficial competition had been retained, and supplies for the industry from the multinationals were both diversified and cheap. One executive in the industry had observed 'we spend more time explaining (the French oil regime) to our corporate headquarters than worrying about it in France'.[22] There had nonetheless been moments of tension. The setting up of CFR with its special rights to the market had been the first, and the tension had been greatest perhaps with the creation of UGP. Not only was there the vexed question of the price of imposed purchases of Algerian oil, but UGP was a new entrant to the market, and as the foreign multinationals feared, the administration of the market subsequently was directed toward fostering the growth of what became the Elf-Aquitaine group. The award of import quotas subsequently was always disputed and this was particularly true of the latest round in 1976.

The most recent award of manufacturing (or import) licenses (A10) had been in July 1973, and there had been some movement in favor of the multinationals, in line with the reintroduction of French Franc Zone crude oil within the quota system (following pressure from the Common Market Commission) and a desire to align the quotas more closely to the market share of the multinationals (50 per cent); 44.5 per cent of the new quotas was given to the international companies. This was still less than they had had before the formation of UGP in 1965, but more than the previous allocation (see Appendix 2). But for A3 licenses, in 1976 for the first time it seemed as if an award of A3 licenses would actually limit the sales of petrol of one company, Mobil, and Esso would at the end of the period covered (up to 1979) reach the limit of its allotment. Not unsurprisingly there was considerable protest,[23] as the structure of prices created by the government was such that petrol sales were the key to company profitability. In overall terms it also appeared that in competing for a market share the maximum share the internationals could achieve was 49 to 51 per cent, while the maximum share the national groups could achieve was 61 to 63 per cent (the variation being due to differences in market growth). As significant was the decision to allow various industries to import heavy crude oil and naphtha, liberalizing the movement of these two products. One effect of this would be to make pricing of oil products within France more influenced by external events and an element of control would thereby be lost. The licenses also gave some help to independents, the companies who had suffered most in the development of the oil policy. Holding a 2 per cent share of the overall market, they were almost exclusively dependent on Elf or CFR for supplies with long-term contracts, and while not strong in petrol held significant shares of other product markets.

In fostering the growth of the national companies, the oil policy had certainly been a success. Although the cost had been considerable in building up Elf-Aquitaine, in encouraging exploration, and supporting IFP (see Appendix 3), there was little evidence that the French consumer had suffered, on balance, thereby. Nonetheless the multinational companies still claimed to be more efficient in bringing oil to the French market. Esso, in 1976, claimed that not only were its oil supplies cheaper than the average for France (average FF388 per ton, Esso FF380 per ton) but also that the internal cash flow of the multinationals, for operations in France alone, was more than twice that for similar operations of the national companies.[24]

The political and administrative control, control of physical planning (like the siting of petrol stations), and other instruments that had been used to encourage the national oil companies had something in common with the direction of other parts of the French economy.

Control of industry was essentially 'dirigiste' and the government had intervened strongly in a variety of other industries considered to be of national importance, notably computers, aircraft, aluminum, automobiles, and steel, with varying degrees of success. The French education system produced an elite which controlled and moved easily between both government and French industry (although not foreign-owned industry). As the senior people had similar backgrounds to those in government-influenced companies, identity of interest and sympathy with policy was more possible than in other countries (except Japan). In addition to this there was a network of interlocking directors, together with common government representatives, sitting on the boards of the French oil companies which inevitably served to keep company policies in harmony with one another and with national policy.

In 1976, however, the success of the policy seemed compromised by external events. The difficulties caused by weak demand, poor capacity utilization, very expensive exploration, and development and price control in the French market seemed to put in question the financial viability of the two national groups, the existence of which could be directly attributed to the policy. With the government under pressure to reduce even further its protection of the market, with the economic recession making continued financial support unattractive, and with a sharp change in energy policy to reduce the large dependence on oil (see below) the future of national oil companies and the oil policy was difficult to foresee. CFP at least had called for direct government support for exploration,[25] but a more long-term development might be the merger of CFP with the new SNEA grouping.

Such a merger would have some notable advantages. In an industry where size is of paramount importance, the combined 1975 sales of $16.3 billion would make such a group the eighth largest world oil company after Standard Oil of California and before Gulf, the tenth largest company in the world,[26] and more than twice as big as the next biggest French company, Renault. In addition it would bring together CFP, long on crude but with compromising obligations to supply the French market, and SNEA, short on crude with no immediate likelihood of reversing the situation. Together, the two companies would have added strength on the competitive international markets. For the immediate future the merger was improbable, not least because SNEA would need several years to adapt to its new structure, even though considerable cooperation in exploration already existed between the two groups.

By 1980 the consistent French preoccupation with supply security took on a renewed urgency. During 1979, crude oil prices rose some 60–65 per cent, partly because of interrupted Iranian oil exports. While France was able to obtain all the crude oil it needed, its oil import bill in 1979 increased by FF18 billion over 1978's bill; in 1980, by FF30 billion. Thus, Prime Minister Raymond Barre stated that France 'must not be delivered gagged and bound into the hands of the oil producers.'[27] It seems that France has come full circle since 1918 and Clemenceau's appeal to Woodrow Wilson for oil!

In answer to the oil 'crisis' of 1979, a 'Hydrocarbon Program' was launched in 1979 with the traditional goals of restricting demand and encouraging supplies of all energy sources. Yet – unusual for the French – the program had several inconsistencies. For one, the government fixed product prices below the levels in France's neighboring countries – hardly a move to restrict consumption. Second, the government imposed an 'exceptional' tax on the depletion allowance, a tax, in fact, that falls more directly on exploration and production than do the so-called windfall profit taxes enacted in the United States in 1980. This tax will surely not encourage exploration, as it hurts the very same companies that have been the most active in exploration. Hardest hit are SNEA, with an estimated incremental tax of FF340 million, and the French affiliate of Esso, at FF125 million.

From exploration efforts begun many years ago, SNEA and Esso coincidentally with the Hydrocarbons Program did find oil in Southwest France. It is estimated that this find will increase domestic French oil production by some 80 per cent; however, domestic production represents a small share of oil consumption. France will continue to have to rely on imports as its major source of oil.

French Energy Consumption

French oil policy can be understood better if one considers oil in the overall energy setting. This setting is characterized by two factors: the heavy reliance of France on imported energy sources, primarily oil but natural gas as well, and the rapidity with which the reliance on the outside world has occurred.

In 1960, domestic energy sources (coal, hydroelectricity, natural gas, and petroleum production in France itself) provided about 70 per cent of French energy requirements. By 1975, with the volume of energy consumption having doubled in fifteen years, imported energy accounted for two-thirds of French consumption. The share of oil in the total doubled in the period under review (see Appendix 4).

Table 2. Oil as a percentage of total energy consumption[28]

		1975	
France	Germany	Italy	United Kingdom
64.6	53.1	69.2	45.0

France's reliance on imported oil was exceeded only by Italy among the major European countries. The reliance of Germany and the United Kingdom on imported oil, while large, was not as large as that of France (and Italy).

France produces small amounts of both natural gas and oil. These amounts are, however, not only small, but declining. In order to secure crude oil under the control of French companies, even if it is outside France, French companies have drastically altered their expenditure patterns in the last ten years. For instance, in 1965 only 40 per cent of total investment was made outside France, while in 1975 this figure rose to two-thirds of all investments (see Appendix 5). Exploration and production investments are being made in such politically 'safe' areas as the North Sea, Italy, Spain, the United States, Canada, and Australia. Of course, investments are continuing to be made in the Persian Gulf, Tunisia, Libya, West Africa, and Indonesia.

Bar some unexpected developments, the heavy reliance of France on imported energy sources will not only continue but will become worse in the foreseeable future. Petrol and gas-oil represent less than 30 per cent of total petroleum consumption. These are products for which, at the present, no substitutes exist. Substitutes do exist, however, for domestic and heavy fuel oils. But coal production is declining and so is that of natural gas. The only viable alternative would be nuclear energy. Thus the French planners in their statement of Plan VII objectives intend that by 1985 the dependency on oil will be reduced by nearly 20 per cent, mostly achieved by an increase in nuclear energy (see Appendix 4) at a predicted cost of FF130 billion. However, for a host of reasons the French nuclear program faces some difficulties and will not be likely to ease the energy problem much in the future.[29] Even with limited success it will have a direct effect on oil company operations, as the biggest impact of the program is to reduce heavy fuel oil consumption. Annual fuel oil consumption is expected to rise from 10.3 million tons in 1975 to 16.6 million tons in 1980 and then drop to 5.1 million by 1985,[30] and this will mean that much refining capacity will have to be expensively converted to produce lighter grades of fuel. The investment required will only be forthcoming if companies operating in France are both profitable and have expectations of so remaining. Before the energy crisis French policy provided these conditions, but its ability to do so in the future is less certain.

Appendix 1. Franc Zone Oil Production and Imports

Franc Zone oil production[31] (thousands of tons)

	France	Algeria	West Africa
1946	51.6	0.2	–
1956	1,263.6	33.5	–
1959	1,610.2	1,232.4	753.3
1960	1,976.5	8,599.7	853.6
1963	2,522.2	23,646.4	998.9
1966	2,932.0	33,257.2	1,509.5

French oil imports[32] (thousands of tons)

	Franc Zone	Rest of world	Finished products
1947	–	5,029	2,102
1956	34	24,979	1,286
1957	122	23,940	2,744
1958	814	27,610	1,749
1959	1,438	27,733	2,125
1960	7,321	23,702	2,687
1961	12,013	23,005	2,462
1962	13,520	23,641	3,070
1963	15,999	27,259	4,047
1964	17,974	31,218	4,415
1965	18,498	40,057	4,068
1966	19,218	43,534	4,916

French oil imports[33] (per cent)

	Near East						North Africa		West Africa					
		comprising												
Year	Total	Saudi Arabia	Iraq	Iran	Kuwait	Abu Dhabi	Algeria	Libya	Congo	Nigeria	United States	USSR	Others	Total
1938	45.0	–	45.0	–	–	–	–	–	–	–	33.3	–	21.7	100
1947	47.7	4.4	22.6	11.8	–	–	–	–	–	–	11.5	–	40.8	100
1950	85.3	23.3	20.8	10.3	25.6	–	–	–	–	–	0.6	–	14.1	100
1955	93.8	11.2	42.9	3.7	30.3	–	0.2	–	–	–	0.2	0.7	5.1	100
1956	91.2	10.3	37.5	6.6	31.5	–	0.1	–	–	–	1.8	0.8	6.1	100
1957	78.5	10.1	18.3	7.4	32.3	–	0.1	–	0.5	–	7.8	0.7	12.4	100
1958	88.3	14.2	29.2	5.3	29.5	–	1.2	–	1.6	–	0.2	0.6	8.1	100
1959	85.1	10.6	31.7	10.2	24.9	–	2.4	–	2.5	–	0.3	0.4	9.3	100
1960	68.2	9.6	25.2	4.4	23.9	–	20.9	–	2.7	–	0.2	0.5	7.5	100
1961	58.2	6.7	19.7	2.3	25.0	–	31.9	0.1	2.4	–	0.1	0.3	7.0	100
1962	55.7	5.6	19.4	4.0	22.1	0.4	34.2	0.7	2.2	–	–	0.2	7.0	100
1963	52.8	4.1	21.4	4.7	18.7	1.1	35.2	3.8	1.8	–	–	0.3	6.1	100
1964	52.8	3.8	16.4	7.0	19.9	3.0	34.8	4.7	1.7	0.6	–	0.2	5.2	100
1965	51.6	4.3	16.8	10.1	15.1	3.1	29.7	10.0	1.9	0.9	–	1.4	4.5	100
1966	48.6	4.9	16.6	6.7	13.3	4.5	29.4	11.5	1.2	2.8	–	2.6	3.9	100
1967	48.1	5.9	18.8	4.4	12.3	3.9	29.9	12.1	0.8	3.0	–	2.3	3.8	100
1968	48.4	5.2	19.6	3.9	9.4	6.5	31.7	14.0	0.6	0.2	–	2.0	3.1	100
1969	44.8	5.6	17.1	4.2	9.4	6.5	29.5	16.9	1.1	2.6	–	2.1	3.0	100
1970	44.1	9.4	12.5	3.7	11.0	5.8	26.9	17.6	1.5	5.3	–	1.4	3.2	100
1971	61.0	16.4	14.9	5.4	10.3	10.5	7.2	14.3	1.6	11.4	–	2.4	2.1	100
1972	66.1	20.6	12.2	6.0	14.5	9.6	9.2	8.3	1.5	11.3	–	1.5	2.1	100
1973	71.5	22.4	13.8	8.0	11.5	11.8	8.2	4.8	1.8	9.3	–	2.5	1.9	100
1974	77.4	34.9	12.9	6.8	9.5	10.1	6.8	2.8	2.8	8.1	–	0.2	1.9	100

Appendix 2

Manufacturing licenses. Evolution of the share between national and international oil companies[34] (per cent)

	A20 1931–1951	A13 1951–1965	A10 1965–1975	A10 1976–1985
International	45.8	50.4	38.7	44.5
National[a]	54.2	49.6	61.3	55.5

[a]Out of which CFR special ← 25.0% →

Market share for all petroleum products[35] (per cent)

	1965		1970		1975	
TOTAL	22.0		24.2		27.2	
Elf	11.5		13.3		14.0	
ANTAR	7.3		8.6		9.0	
Other	5.8		3.0		2.5	
National oil companies		46.6		49.1		52.7
Shell	18.5		18.1		15.6	
Esso	14.2		12.7		12.1	
BP	11.7		11.7		11.0	
Mobil	5.6		5.3		4.9	
Fina	3.4		3.1		2.7	
Other					1.0	
Multinationals		53.4		50.9		47.3
Total		100.0		100.0		100.0

Appendix 3

'Fonds de Soutien' sales tax proceeds to BRP/ERAP[36]
(millions of francs)

Per annum receipts		Total receipts	
1951	95[a]	1951–59	1,286[a]
1959	190	1960–73	4,356
1961	213		
1963	200		5,642
1965	221		
1967	363		
1969	288		
1971	200		
1973	200		

[a]*Note:* These figures are estimates based on market size.

Sales tax 'IFP'[37] (millions of francs)

	Exploration and exploitation	Refining and petrochemicals
1960	11.2	11.1
1965	35.0	31.2
1970	59.1	65.2
1971	58.5	70.9
1972	64.4	69.3
1973	74.1	77.1
1974	82.3	88.6

Total receipts	
1960–65	267.5[b]
1966–70	505.5[b]
1971–74	685.2
	1,458.2

[b]Estimate based on interpolation.

Appendix 4. French Energy

Pattern of French energy consumption[38] (per cent)

	1960	1970	1975
Coal	54.7	25.6	16.7
Hydroelectricity	10.4	8.3	8.2
Nuclear energy	–	0.8	2.6
Natural gas	3.5	6.5[a]	10.0[a]
Petroleum	31.4	58.8	62.5

[a]Includes imports, primarily from the Netherlands. In 1975, for instance, nearly 40 per cent of the natural gas consumed had been imported.

French consumption of major petroleum products, 1975[39]

	Million tons	Per cent
Petrol	15.9	19.6
Gas oil	6.9	8.5
Domestic fuel oil	29.6	36.4
Heavy fuel oil	28.8	35.5
	81.2	100.0

French energy consumption, 1975–85[40]

	1975		1985	
	Per cent	Million tons oil equivalent	Per cent	Million tons oil equivalent
Oil	62.5	102	42.2	98
Coal	16.7	28	10.8	25
Gas	10.0	17	16.0	37
Nuclear	2.6	4	13.7	55
Hydro	8.2	13	6.0	14
Geothermal/solar	–	–	1.3	3

Appendix 5. French Oil and Gas Production. French Oil Industry Investment

Production of natural gas in France, 1965, 1970, 1974, 1975[41] (billions of cubic meters)

	1965	1970	1974	1975
	5.1	7.0	7.6	7.4

Production of crude petroleum in France, 1965, 1970, 1974, 1975 (millions of tons)

	1965	1970	1974	1975
	3.0	2.3	1.1	1.0

Pattern of French oil industry investment, 1965, 1970, 1974, 1975[42] (per cent)

	1965	1970	1974	1975
In France	60.0	61.7	44.9	33.4
Outside France	40.0	38.3	55.1	66.6

History

1917 Dec. Petrol crisis (complete shortage foreseen March 1918).

1918 August Policy group formed.

1920–28 Fierce Esso/Shell competition – free imports.

1920 Société Alsacienne des Oléonaplates established. Later (1927) called SOCANTAR.

1921 France takes 25 per cent share in Turkish Petroleum Company (later IPC).

1924 March CFP (Compagnie Française des Pétroles) formed (the state had rights to 90 per cent production and superprofits) with

share in IPC of	23.75%
also	
Anglo-Iranian	23.75%
Shell	23.75%
Standard of New Jersey and New York	23.75%
Gulbenkian	5.00%

1925 Jan. Organisation Nationale des Combustibles Liquides (ONCL) formed for research and exploration in France and abroad. Ecole du Pétrole established at Strasbourg University.

1928 March Import controls with quotas; two systems

– import of finished products, A3 customer system for providing protection for refining

– import of crude, A20 import system protecting indigenous operators.

obligations – stocks
– French management
– 66 per cent French flag
– undertaking, when required, of contracts of national interest.

1929 First A3 awarded.

1930 June CFR (Compagnie Française de Raffinage) formed. CFP holds (1975) 56.53 per cent. Given right to refine 25 per cent of demand.

1931 March State takes 35 per cent of CFP shares and 40 per cent of voting. French issue of A20. French groups receive 54.2 per cent.

1933 ANTAR created. Formerly Raffineries Pechelbronn et Serco.

1937 Creation of Centres de Recherches de Pétrole du Midi (CRPM) by ONCL.

1939 Creation of RAP (Régie Autonome de Pétrole) for gas exploitation in Aquitaine, Saint Mercet field, discovered by CRPM.

1941 SNPA (Société Nationale des Pétroles d'Aquitaine) created to explore and exploit Aquitaine oil.

1943 Nov. IFP (Institut Français de Pétrole) created with petroleum product sales tax for research and training.

1944 SNPLM (Société Nationale des Pétroles du Languedoc Mediterranéan) created to exploit Alsace and Languedoc oilfields and explore.

1945 Oct. Creation of BRP (Bureau de Recherches de Pétrole) to carry out Government policy on
– research
– French majority companies
– infrastructure } related to oil
– capital investment }
in conjunction with public, private, and mixed firms like SNPA, SNPLM, and associated groups in Algeria and West Africa.

1947 DICA (Direction des Carburants) of the Ministry of Industry and Commerce given control of Ecole Nationale Supérieure du Pétrole and financed by IFP.

1949 SNPA finds oil at Lacq Supérieur.

1950 A13 crude import licenses issued, French groups take 49.6 per cent. 'Fonds de Soutien' – sales tax – introduced and used principally to fund BRP.

1951–53 SNPA finds gas at Lacq Profond.

1954 Esso finds oil in Les Landes.

1955 Reformulation of exploration rights covering exploitation rights. Restructuring of CFP marketing companies to form TOTAL. SNPLM changes to Compagnie d'Exploration Pétrolière (CEP).

1956 Oil strikes in Algeria and Gabon.

1956 Oil strike in the Congo.

1957 SOGERAP (Société de Gestion des Participations de la Régie Autonome des Pétroles) – to manage the RAP portfolio.

1959 TECHNIP, engineering firm, formed by IFP. First Algerian ton of crude oil sold on market.

1960 UGP (Union Générale des Pétroles) created to refine and market the production of
- RAP
- SNRepal
- BRP subsidiaries.

UIP (Union Industrielle de Pétrole) formed for Ambès (Bordeaux) refinery
- 60 per cent UGP
- 40 per cent Caltex.

1961 Imports of crude ten times 1929 amounts at 30 million tons crude plus 2.2 million tons finished products. Crude from French Zone freed from import control.

1963 A10 crude import licenses issued – for the first time tonnage lower than market, obliging the purchase of French crude. 61.3 per cent goes to French companies.

1965 Tax concessions to French-based companies. A3 issued for new (refined) products. French companies receive 52.85 per cent.

1965 ERAP (Establissement de Recherche et d'Activités Pétrolières) created, by fusing BRP and RAP.

1967 UGP becomes Elf Union. Elf mark created and ERAP marketing companies restructured.

1970 Compagnie de Pétrochimie formed, 50 per cent SNPA–ERAP and 50 per cent TOTAL. SOCANTAR acquired by Elf–ERAP (51 per cent).

1971 Feb. Algerian assets of CFP and Elf nationalized.

1973 Oil crisis. Schvartz report on oil company practices.

1974 SNPA invests in Le Nickel.

1974 Feb. Protocol on permitted commercial practice signed by oil companies and government.

1976 SNEA (Société Nationale Elf-Aquitaine) created. Memorandum to Brussels Commission by national oil companies on controlling refining capacity.

1979 France suffers no supply disruptions due to the suspension of Iranian oil exports. However, the increase in the (imported) crude oil prices affected the French balance of payments adversely.

1980 Elf/Esso discover an oilfield in the Aquitaine Basin, doubling France's (modest) domestic crude production.

References

1. La Politique Pétrolière Française, Professeur M. Devaux-Charbonnel, *Cours de Droit Pétrolier*, Paris I et II.

2. La Politique Pétrolière Française, Professeur M. Devaux-Charbonnel, *Cours de Droit Pétrolier*, Paris I et II.
3. *L'Intervention de l'Etat dans le Secteur Pétrolier en France*, S. Murat.
4. *Legal Organization of the Oil Industry*, Shell Oil Company.
5. *Bloc-Notes*, économique no. 5/68, December 1963, Shell.
6. Richard F. Knisel, *Ernest Mercier*, University of California Press, 1967, pp. 33–44.
7. *The Oil Era, TOTAL in the Oil Industry*, CFP Publication, 1966, p. 5.
8. *The Oil Era, TOTAL in the Oil Industry*, CFP Publication, 1966, pp. 11–12.
9. *The Oil Era, TOTAL in the Oil Industry*, CFP publication, 1966, pp. 11–12.
10. M. A. Adleman, *The World Petroleum Market*, Johns Hopkins Press, Baltimore, 1972, p. 137.
11. *Ordonnance 45-1483 du 12.10.1945.*
12. *Legal Organization of the Oil Industry*, Shell Oil Company, p. 5.
13. *Bloc-Notes*, économique no. 5/68, December 1963, Shell.
14. Pétrole 74, *Activité de l'Industrie Pétrolière*, CPP, 1974, p. 101.
15. *Elf Group Pamphlet*, 1971.
16. *Report on Refinery Details and Policies of European Countries*, EEC publication.
17. 'The Changing Scene for Oil', *Petroleum Economist*, May 1976, p. 168.
18. Private communication to the author, 1976.
19. Private communication to the author, 1976.
20. *CFP's Annual Report 1975*, Elf-Aquitaine 1975, BP Statistical Review.
21. *BIP*, March 29, 1976.
22. Spokesman for oil MNC French affiliate to the author, 1977.
23. 'Mobil Protests Action by France', *New York Times*, October 6, 1976.
24. *Esso Press Conference*, March 10, 1977, by M. H. Lamaison, President Esso (France).
25. *CFP 1976 Annual Report*, p. 24.
26. *Fortune*, August 1976.
27. *Petroleum Economist*, November 1979, p. 477.
28. *BP Statistical Review of World Oil Industry, 1975*, BP, London, 1976, p. 16.
29. *Petroleum Economist*, May 1976.
30. *Le Raffineur Français*, CFP 1977, p. 29, from VIIe Plan.
31. *Activité de l'Industrie Pétrolière*, Comité Professionnel de Pétrole (CPP), 1965, B15 and B31.
32. *Activité de l'Industrie Petrolière*, Comité Professionnel de Pétrole (CPP), 1966, B15 and B31.
33. *Activité de l'Industrie Petrolière*, Comité Professionnel de Pétrole (CPP), 1974, B101.
34. Private communication to author, 1977.
35. *BIP*, March 29, 1976.
36. *BIP*, March 29, 1976, CPP, p. 301.
37. *BIP*, March 29, 1976, p. 302.
38. *L'Industrie Française du Pétrole, 1975*, Union des Chambres Syndicales de l'Industrie du Pétrole, Paris, 1976, p. 35 and p. 69.
39. *L'Industrie Française du Pétrole, 1975*, Union des Chambres Syndicales de l'Industrie du Pétrole, Paris, 1976, p. 8.
40. *Petroleum Economist*, May 1976.
41. *L'Industrie Française du Pétrole, 1975*, Union des Chambres Syndicales de l'Industrie du Pétrole, Paris, 1976, p. 27 and p. 11.
42. *L'Industrie Française du Pétrole, 1975*, Union des Chambres Syndicales de l'Industrie du Pétrole, Paris, 1976, p. 32.

CHAPTER 3

Compagnie Française des Pétroles

Compagnie Française des Pétroles (CFP) is the largest company in France. Its sales of $9.15 billion in 1975 made it the biggest European national oil company. ENI of Italy was second with $8.3 billion of sales.[1] CFP was distinguished from the other French national oil company, Société Nationale Elf-Aquitaine (SNEA) by being older, having a larger percentage of private shareholding, crude supplies in excess of its requirements, and a substantial international marketing network. Nonetheless it was controlled by the French state and as such was an instrument of French oil policy with special obligations obviously not shared by the affiliates of the oil multinationals which also operated in France. In combining the interests of the state and private investors, in having specific national objectives, and in being a significant international operator, CFP was the embodiment of the contradictions and successes of French oil policy.

The Creation of an Oil Company

According to Senator Henry Bérenger, the president of the Comité Général du Pétrole which was set up to deal with the severe oil-supply difficulties experienced by France during World War I, 'The . . . victory was . . . won by the blood of the Poilus,* the Tommies, the Arditi,† and the Yanks, but it could not have been won without that other blood produced by the earth called oil.'[2] His opinion was widely shared and the French government was determined thereafter to obtain a more specific interest in oil affairs than it had had before the war.

As a result of the San Remo treaty of 1920, which dealt with the confiscation of Germany's foreign assets, the French government acquired the rights to shares formerly held by the Deutsche Bank in the Turkish Petroleum Company (TPC), an exploration company with interests in Iraq. Initially TPC united the interests of Shell, Anglo-Persian (later British Petroleum), and Mr. Gulbenkian, but the admittance of France led to a long argument with the Americans; eventually Standard Oil of New York (now Mobil) and Standard Oil of New Jersey (now Exxon or Esso) joined the group. While these arguments were being resolved, the French Prime Minister, Raymond Poincaré, was deciding how to take up the shares in TPC. In 1923, he asked a well-known industrialist, Ernest Mercier, who

*Poilus: bearded ones, the French nickname for their troops.
†Arditi: Italians.

already had some oil industry experience, to form a private company uniting the interests of the petroleum distributors in France and the major French banks and which would follow specifically French interests as Anglo-Persian followed British interests. Private investors were at first reluctant to participate. They saw the new company as the first step to a state monopoly, and they were unwilling to undertake what might be an expensive venture with great risks (despite the promising appearance of the Iraqi concession); but by 1924 Compagnie Française des Pétroles was established. Most of the shareholders were French companies but subsidiaries of foreign oil companies bought sufficient CFP stock to be represented on the board as well.

The prospect of a left wing government coming to power in the 1924 elections led Mercier to conclude with the government of Poincaré a hurried agreement defining the rights of CFP in exchange for the TPC shares. The government had the right to name two commissioners (with limited powers) to the board of directors; the chief executive officer and directors were to be approved by the government; the state would receive a share of excess profits, and would have a general option to purchase 80 per cent of the company's oil, an option that was subsequently restricted by the incoming left wing government to defense requirements only. If this agreement served to define the role of CFP, that of TPC was not at all clear. There was disagreement as to whether the company should operate on its own, and provide its shareholders with dividends, or whether it should be limited to crude oil production which would be supplied to the individual partners for their own uses. Mercier, who intended building up a French refining industry, strongly favored the latter, as did Standard Oil of New Jersey. The argument came to an abrupt end on October 15, 1927, when oil was struck in the Mosul field in Iraq, and the individual partners agreed to take oil and not dividends. They also finally determined the size of their individual shareholdings in TPC including Gulbenkian's famous 5 per cent. The agreement was finalized in July 1928, TPC renamed itself Iraq Petroleum Company (IPC), and its shares were held as follows:

Compagnie Française des Pétroles	23.75%
Royal Dutch/Shell	23.75%
Anglo-Persian (BP)	23.75%
Standard Oil of New Jersey (Exxon) Standard Oil of New York (Mobil)	23.75%
Mr. Gulbenkian	5.00%

With a certain supply of oil, and renewed private interest in CFP, following the oil strike, Mercier turned his mind to establishing an outlet for the oil by creating a refining company. A law passed in 1928 (see Chapter 2) gave the government control of French oil imports and encouraged the formation of a French refining industry. However, the affiliates of foreign oil companies on the board of CFP were strongly in opposition to the idea of CFP having its own refining company, and they were supported by some of the French distributors who feared the potential monopoly power of CFP. Turning to the

Prime Minister, the re-elected Poincaré, Mercier concluded an agreement in March 1929 whereby CFP would be transformed into a mixed company, with the government holding 25 per cent of the shares and appointing one quarter of the directors. CFP was also authorized to create a refining subsidiary, Compagnie Française de Raffinage (CFR), in which the government would hold 10 per cent of the shares and CFP 55 per cent. Most controversially, petroleum marketers would be obliged to take 25 per cent of their requirements from CFR.

Financial participation by the state required parliamentary approval and a long wrangle over the agreement ensued with both Left and Right raising objections to the new form of state intervention. Ratification was finally achieved on March 4, 1931, with some changes to the original agreement. To accommodate the objections of the independent refiners, the 25 per cent clause was revised to give CFR the right to refine 25 per cent of the nation's needs. To accommodate the objections of the Left, the government's stake in CFP was raised to 35 per cent of the shares and 40 per cent of the votes, the state's portion of excess profits was increased, and the right of the state to purchase 80 per cent of CFP's oil production was revised again from purely defense needs to cover all state requirements; this right has never been exercised.[3] Henceforth, CFP and CFR worked within these parameters; CFP was to develop French oil interests in the Middle East, CFR was to develop the refining industry in the protected French market.

The remaining requirement for CFP to fill Mercier's ambitions of creating an integrated company was for a link between the oil production of CFP and the envisaged refineries of CFR. CFP was already involved with its IPC partners in building a pipeline from Iraq to the Mediterranean, a difficult technical problem aggravated by differences between the French and the British concerning the route. IPC represented a crucible for the pre-World War II power struggles and each country wanted the pipeline to pass through a country under its influence. Eventually two branches, one to Haifa and one to Tripoli, were built. In 1931, Mercier founded Compagnie Navale des Pétroles (CNP) to carry the oil and in August 1934, the first cargo of French oil produced in Iraq reached CFR's new refinery at Le Havre.

Construction, Destruction, and Reconstruction

Through the thirties, despite the difficult economic conditions in France, both CFP and CFR grew steadily. CFP concerned itself with oil exploration to satisfy the fast-growing needs of CFR, which could and did absorb more oil than CFP could supply. Additional supplies came principally from Venezuela and the United States. Initial effort by CFP was directed to the Mosul fields, but additional concessions were also taken in partnership with the other members of IPC. Hence CFP found itself working in South Iraq, Qatar, and on the Trucial Coast. In Kuwait, CFP formed an exploration company with Anglo-Persian and at the end of the decade the various investments began to pay off as oil was struck in Kuwait (1938) and Qatar (1940).

Delivery of oil to France continued to be made by Compagnie Navale des

Pétroles, which ordered its first tanker in 1935, the year in which CFP opened its second refinery, at La Mède. Supplies were also provided by another shipping company, Compagnie Auxiliaire de Navigation (CAN) which was one of the founder shareholders of CFP. Crude taken from CFP by CFR was bought f.o.b. and shipped by either or both of these two shipping companies, thereby not only fulfilling one requirement of the 1928 petroleum import law which required that two-thirds of French oil imports were to be in French flagships, but also thereby underlining the distinction between CFP and CFR. The distinction arose from the fact that CFR's role was seen by its owners as being exclusively in the special situation of the French market, where it was restricted to refining and some few closely defined marketing operations.

CFR was constrained in its marketing by agreements with its private shareholders, most importantly the independent distributors who, by virtue of their special relationship with CFR, were known as 'adhérents de la CFR'. The most important of these were Desmaris Frères, Lille Bonnières et Colombes, La Société Française des Carburants, and l'Omnium Français de Pétrole. The agreements with the 'adhérents' and the 'confrères' – companies who were exclusively supplied by CFR – were complex and so restrictive that CFR was not allowed to market its own products except in the special case of very large (and not very numerous) fuel oil users such as the cement, car assembly, steel, and electricity companies. Outside France product marketing was done by CFR on a short-term basis on the open market, generally for small quantities to balance production. Sales reached as far afield as Finland (gas oil) and Australia (petrol) intermittently. But the dominant preoccupation of the company through its first decade was with building up and perfecting its refining capacity. By 1936, CFR was supplying nearly 20 per cent of French demand[4] from its two refineries, with IPC providing nearly 42 per cent of all crude imports.

The only major marketing initiative was taken in 1933 when both CFP and CFR (50 per cent each) invested in a bunkering company, Société des Pétroles Mory. For many years this was to be the only marketing initiative of both companies and by the nature of its business Société Mory did not have much impact on the internal French market. By the time that World War II started, Mercier was the leader of a successful and vertically integrated oil company which lacked the final link to the consumer, marketing of finished products. By finding oil so early in its history, and by virtue of the controlled and profitable market in France, CFP and CFR had managed to expand and invest without any further government support than subscription to the capital of CFP and CFR to keep the share of the state to 35 per cent and 10 per cent respectively. However, Mercier, out of favor with the Vichy regime, resigned from CFP in 1940 and his successor, Jules Meny, was deported to Dachau and did not survive. The Middle East interests of CFP were cut off by the Allies, but IPC continued exploring and CFP's interests were guarded by representatives in London.

At the end of the war a new president, Victor de Metz, was appointed, who, in contrast to Mercier, would have nothing to do with the press and had little use for government agencies. The immediate objective of de Metz was to re-establish

CFP in the eyes of its IPC partners as an equal. To do this, he had to rebuild the organization of CFP and the much-damaged refineries of CFR. Re-establishing the credibility of CFP was no easy matter. The other partners in IPC were not above raising the capital of IPC in the hope that CFP would not be able to subscribe.[5] However, CFP was able to undertake its full partnership obligations and the exploration and production activities in Iraq continued to expand to its benefit. In France, two developments took place which could have affected the rebuilding of CFP. One was the establishment by General de Gaulle of a public corporation to look for oil on French territory for the exclusive use of the nation (a stronger commitment than that of CFP), and the other development was the program of nationalization of the post-war government. CFP remained on the periphery of these events for interrelated reasons. It had significant private shareholdings and many international relationships which would have made an exclusively national role impossible and nationalization difficult, and it was also preoccupied with re-establishing itself rather than undertaking new tasks. Thus, while de Metz gave his encouragement to the new exploration company (Bureau de Recherche des Pétroles – BRP), he was left alone to rebuild CFP.

In 1946, CFP took a majority holding in a distribution company in French West Africa and thereby established its first distribution outlet outside France. Then, foreseeing that its supply obligations in France would be difficult to meet with only its proportional share from the IPC holdings, CFP negotiated a 'Heads of Agreement' whereby CFP received an increased off-take from its IPC partners.[6] In 1948, substantial oil reserves were found in Southern Iraq and in the same year CFP undertook with its French partner (CRP) an exploratory mission to the Algerian Sahara, its first venture outside the Middle East. In France, CFP's principal subsidiary CFR was repairing war damage and preparing itself for the projected expansion in demand; its relationships with the distribution companies remained the same as they had been before the war. In 1950 when the import controls established in 1928 were re-established by the French government following their rupture by the war, observers of both companies might have foreseen a return to the *status quo ante* following the disruption of the war. CFP would be active in the Middle East as a producer and supplier to CFR and CFR would be an important refiner in France. Neither company would be greatly involved in marketing as CFR could absorb all CFP's production, and CFR was held back from entering the marketing phase of the business by the independent distributors.

CFP as a Market Force

As noted above, de Metz was an uncommunicative man and he did not publicly spell out his strategy for CFP. However, two principal lines of thought soon became clear. While continuing to fulfill the obligations to CFR, de Metz had every intention of making CFP a notable presence on the world oil scene, and of running CFP like a multinational oil company. Second, CFR would be left to operate in France as a company to which CFP would act as a crude supplier. Two facts made this policy workable: one was that CFP had every prospect of becoming long on crude following the discoveries made just before and after the war despite CFR's growing requirements; the other was that CFR was acting in what was to all intents and purposes a protected market and could therefore be left to work out its own destiny. In this way CFR's status was comparable to the refining and distributing subsidiaries of the oil multinationals in France except that CFR had a right to 25 per cent of refining. As it turned out, in the area in which both companies took initiatives in the fifties, namely marketing, their coordination efforts were, in fact, quite strong.

The controls imposed on the French market did not prevent vigorous competition within it, and in the early fifties the independent distributors began to experience difficulty in competing with the well-organized marketing campaigns of the affiliates of the foreign multinationals. The government foresaw in this a repeat of the conditions that had existed before and after World War I, when French refiners and distributors had had difficulty staying in existence. A study was therefore initiated under the auspices of CFR which recommended an improved marketing approach: in return for accepting certain standard procedures the independent marketers could use a new brand name common to all affiliates of CFR which would give them greater weight in the market place. Hence in December 1953, the brand name TOTAL was created. It was used by CFP's distribution in West Africa from the end of 1953 onwards, and appeared in France in 1957. Although some of the bigger independents kept their own brand names, twenty independent distributors started selling under the TOTAL name.[7] The appearance of TOTAL coincided with the crude production of CFP exceeding the requirements of CFR; thereafter CFP always had more crude than CFR could absorb.

CFP's oil production had increased steadily. In 1949 the Qatar discovery started producing and in 1952 the new discoveries in Iraq also started to flow. In the same year, to consolidate the exploratory mission undertaken in 1948, CFP set up its own exploration company in Algeria which was to work in partnership with SNRepal, a joint investment by BRP and the Algerian government. It was CFP's first company outside the orbit of its IPC partners. CFP also invested in 1952 in an existing refinery in Lisbon, and two years later it invested in another refinery in Trieste. In 1954, CFP obtained a share in the new Iran Consortium which was formed following the nationalization of Anglo-Persian's assets in Iran and included the Abadan refinery. The share that CFP took, 6 per cent, was to provide it both with products and crude that were to be an important contribution to its operations. To accompany the growing refinery operations outside France, CFP began to enlarge its marketing operations under the new brand name TOTAL. Through the rest of the decade, CFP expanded its West African operations, moved into Australia, East, South, and North Africa (with the exception of Algeria where marketing was undertaken by CFR), built up its operations in Portugal and Italy and started to operate in Germany. More oil discoveries were to accompany this expansion. In 1956 a large oil, and larger gas, find was made in the Sahara and the oil came on stream two years later. These discoveries were important, being the first substantial finds in the Franc Zone, because they gave a boost to French confidence and to CFP's private investors who continued to support the company. In 1959, another large discovery was made in Abu Dhabi onshore and a year later there was another strike offshore. CFP thus entered the sixties with a fast-growing marketing network of its own and with oil supplies more than equal to its requirements. Since 1952, production had gone up almost four times from 4.9 million tons in 1952 to 19.1 million tons in 1960.[8]

In France, CFR continued to expand its refinery operations, particularly at Gonfreville in Normandy, but its customers, the independent distributors, were having a less happy time. The appearance of the brand name TOTAL was not timely enough to prevent some of the smaller distributors from falling into financial difficulty. In 1955, CFR took an interest in two companies: together with BP in La Société Parisienne des Essences and alone in l'Union des Pétroles. The latter was renamed Compagnie Française de Distribution – TOTAL and was the first direct outlet that CFP, through CFR, had to the French market. Continuing fierce competition squeezed the independents and in 1960 several other distributors (either 'adhérents' or 'confrères' of CFR) were taken over and CFR's market share rose to 12 per cent. One distributor of particular importance that was taken over by CFP was l'Omnium Français de Pétroles. This acquisition gave CFP distribution facilities in Morocco, Algeria, Tunisia, and Lebanon.

The sixties saw no slow-down in the development of CFP as both its production and its markets expanded ever more. Production of crude oil by CFP had slightly exceeded its requirements since the early fifties by some 10 per cent but rose to 50 per cent of requirements by 1960, at which level it stabilized (with rising requirements) through the next decade. By 1965 oil production had nearly doubled from 1960, reaching 37.8 million tons, and CFP was responsible for 6 per cent of Middle East production and 25 per cent of Algerian production[9] (see Appendix 1). New strikes had been made in the Sahara, and exploration was pressing forward everywhere. The shipping fleet of CNP was four times its 1950 tonnage in 1965. Further refinery investments had been made both in France and outside and the markets everywhere were expanding vigorously. In 1966, the biggest of the French independent distributors, Desmaris Frères with 10 per cent of the market, was taken over by CFR, bringing its market share to 22 per cent and prompting it to change the name of the distribution company to its present form TOTAL–Compagnie Française de Distribution (TOTAL–CFD). There were still (and are still) twenty-one other distribution companies who retained a special relationship with CFR (and in some cases were joint investors with CFR and private investors) distributing CFR products with or without the TOTAL brand name.

A Partner and Rival Emerges

The second half of the sixties brought a change to CFP's position in France which it did not welcome unquestioningly. De Metz had from the inception of BRP encouraged its specific efforts of exploring for oil in the Franc Zone and had himself taken initiatives in the francophone area, notably in Algeria (and in West Africa). As the company most dominant on the French market, especially since the absorbtion of Desmarais Frères, in no need of government funding due to the surplus crude situation, and being more preoccupied with finding markets than finding oil, CFP had no need to object to the activities of BRP and its sister exploration companies RAP (Régie Autonome de Pétrole) and SPAFE (Société du Pétrole d'Afrique Equatoriale). But the discovery of oil in large quantities in Algeria and Gabon and the desire of the French government to find an outlet for it changed the relationships. Algerian and West African crude was more expensive than Middle East crude, and the market was anyway a buyers' market. So when the government created a new refining and distributing company, Union Générale des Pétroles (UGP), to absorb the production of BRP, RAP, SPAFE, and their affiliates, and imposed on companies operating in France an obligation ('devoir national') to take a certain percentage of their crude supplies from these companies, CFP's voice was to be heard among the objectors.

Although the 'devoir national' obligations were reduced by an amount related to the activity of any company in exploring and producing in the French territories, they were nonetheless an increase in the crude supply costs of companies like CFP and its subsidiary CFR. Also the price of the oil was considered artificially high, and therefore represented an indirect subsidy to the state companies. The price was fixed between the transfer price of crude sellers to their French affiliates and the price of marginal crude sales on the open market. As CFP was not alone in charging a high transfer price, the price of 'national' oil was an astute squeeze on the oil multinationals and CFP profits. Important too for CFP was the emergence of a new force in the market, UGP; whatever the ties of common nationality it was a new competitor and directly government supported. Reputedly the attitude of CFP to Franc Zone oil prices earned de Metz the lasting enmity of President de Gaulle and Pierre Guillaumat, the former president of BRP and subsequently the first Chief Executive Officer of Elf-Aquitaine. But if this was true it was also true that de Metz now controlled a company of such size and ramifications as to make himself a considerable economic and political presence in his own right and well able to meet any criticism. As often in the oil industry, disagreement on one front was not carried to the other fields of activity and CFP continued to cooperate and work with the French exploration companies, particularly in the North Sea where exploration activity was fast increasing.

Also CFP was moving into North America; exploration started in Canada in 1956 and ten years later CFP had bought into a refining operation in Michigan, giving it access to the American market. The appearance of the new national rival, which in 1967 took the Elf trademark, did not stop the growth of CFP, which had been steady since the middle fifties at home and abroad. In addition to growing refining and distribution activities everywhere, CFP had begun to diversify in France. A by-product of CFR's refining activities had been, since the establishment of a cracking unit at the La Mède refinery in 1950, various petrochemical products. From 1959 large quantities of ethylene were produced from the Gonfreville refinery and by 1968 petrochemical investment was sufficiently important to figure separately in the balance sheet (being previously consolidated with refining). At this time 'other products' constituted 6.5 per cent of sales. CFP's main petrochemical activities were controlled by a subsidiary TOTAL Chemie in which CFR held half the capital. In 1969 in response to Government pressure to strengthen French petrochemicals, TOTAL Chemie went into equal partnership with another French company, Société Nationale des Pétroles d'Aquitaine (SNPA) in which the state had an interest via the Elf group, to form the petrochemical company ATO Chemie which regrouped the interests of both its partners into the single company.

By the end of 1970 the international oil scene was showing signs of losing the stability which had been its characteristic for many past years and in early 1971 CFP's (and Elf's) assets in Algeria were nationalized. For France the event was a shock in view of the importance given to Algerian production, and this was particularly the case for the Elf group. For CFP the Algerian production was less important and an agreement was reached with the Algerian government relatively quickly, an agreement made easier by the simpler company structure under which CFP was operating (i.e. no complicated cross-holdings or partnerships).

While these events were taking place, de Metz retired from the president's position at CFP, taking the post of 'Président d'Honneur'. He had led CFP for a quarter of a century during which time CFP had reconstructed itself from almost nothing to become the largest integrated non-Anglo-Saxon oil company; in the decade of the sixties alone the growth had been remarkable. At the base of the progress of CFP was its oil production which had risen 30 per cent more than world production, since 1960, from approximately 20 million tons to 60 million tons, and which had consistently been 50 per cent more than requirements (see Appendix 1). Despite the production in Algeria, the amount supplied from CFP's Middle East fields had fallen only slightly from 80 to 70 per cent.

To dispose of so prolific a production, the emphasis of the group had continued to move away from an accent on upstream activities to investments in downstream activities; and in the period 1961 to 1970 the net assets in exploration and production had dropped from 45 to 25 per cent of the total, while refining and distribution activities had climbed from 45 per cent to 65 per cent of the total. The largest expansion had been in markets outside France where expansion had been 30 per cent greater than in France and where sales of finished products were just under half the group total (1970 sales in France, 21,497 million tons; outside France, 19,050 million tons).[10] The group was then a truly multinational operator, with markets outside France in Britain, Italy, Germany, and Portugal, most of Africa (Egypt and Zaire being the only significant exceptions), Australia, and the United States, and production in the Middle East, North Africa, and Canada. This achievement was in agreement with the original objectives of de Metz when he had taken over the war-decimated group. His one regret was that the price earnings ratio of CFP's shares was still not equal to that of majors operating in France. That the international network of CFP was very much in accord with that part of French oil policy which sought as strong a presence overseas by French companies as foreign companies held in France was almost an accident. The crude-long company had been obliged to follow this path since, because of French oil policy with its accent on competition and the peculiar restraints on CFR in its early history, it could not take over the whole French market. Nonetheless CFP still had special obligations to the French government, which its international role obscured.

CFP in the Seventies

The new president of CFP, René Grenier de Lilliac, came originally from the French government's Direction des Carburants (DICA), was the vice-president of CFP, and had been president of CFR for five years. He took office in 1971 and was more open and responsive both to his own executives and to the press than de Metz. His most immediate problem was to deal with disruption of deliveries due to the nationalization in Algeria (through increased production in the Middle East). 1971 brought new successes for the combined CFP–Elf North Sea exploration group, Petronord, in which CFP held a 24 per cent share. A large gas find, the Frigg field, was struck and a smaller structure, Heimdal, looked promising (although in 1976 its development was postponed indefinitely). Also oil production started in Ekofisk, in which CFP had a small interest (4 per cent); all three fields were in the Norwegian North Sea. But despite encouraging signs, the task of the president now appeared to be more complex than that of de Metz (who had retired at the point which was subsequently seen as the end of the boom period of the Western world through the sixties). The rising cost of crude oil, the rising cost of exploration and development in areas such as the North Sea, and the accelerating worldwide inflation were serving to squeeze self-financing margins for the oil companies, and in his first annual report for 1971 Grenier de Lilliac was already foreseeing financing difficulties; the share price of CFP showed signs of weakening (it had for a long time been stronger than the average market index).

In 1972, IPC was nationalized, but nationalization did not extend to CFP's other company, in Iraq, Basrah Petroleum Company, which operated in the south of the country. Astute diplomacy by the French government and CFP averted the disruption this might have caused and a ten-year agreement was made with the Iraqi government which gave CFP special treatment in relation to dispositions made by IPC as a result of the nationalization. In the same year, as a member of the Iranian consortium, CFP was party to a twenty-year agreement with Iran, more favorable to the latter than preceding agreements. The energy crisis of 1973, the quadrupling of crude oil prices, the subsequent downturn in worldwide economic activity, and the demand for oil were to complete the change for the worse in CFP's fortunes.

Difficulties in the tanker market in 1974 due to reduced demand led CFP, through l'Omnium Français des Pétroles, to take over Compagnie Auxiliaire de Navigation (CAN) in which it had taken a 24 per cent stake in 1970. Together with CNP, a CFP company, CAN did the bulk of the oil transport for CFR, in compliance with the French flag requirements of the government (see the section 'Construction, Destruction, and Reconstruction') and both CNP and CAP carried for CFP for delivery outside France when they could price competitively. The acquisition took the fleet under the control of CFP from 14.8 million tons to

24.8 million tons.[11] Reduced demand was accompanied by reduced production. In 1975, production was for the first time less than the previous year – equivalent to 1972 production and 17 per cent less than 1974 (Appendix 2). And, of course, the reduced demand was reflected by reduced use of refinery capacity; in 1975 CFR's refineries were working at only 67 per cent capacity. The situation was worse in France than in CFP's overseas markets where sales had held up remarkably well, particularly in Britain where CFP had bought out an American independent, ARCO, in 1974. CFP like other oil companies was caught by the slowing down in the growth of demand which the energy crisis had provoked. In 1971 refinery utilization had been 90 per cent (capacity was 17 million tons outside France, 30 million tons in France). By 1973, utilization was down to 86 per cent following an increase in French capacity to 40 million tons and in 1974 it was down to 80 per cent following refinery expansion in Italy, Holland, and France (bringing the capacity outside France to 26.7 million tons, the French capacity to 43 million tons)[12] (Appendix 3). With controlled prices in France and Italy, a fierce price war in Germany, reduced demand everywhere, and much less preferential crude due to nationalization by the producers, CFP found itself in a cash squeeze.

One of the reasons why the refining situation of the whole CFP group was unbalanced was that the distribution of refining capacity did not correspond to the distribution of sales. For some years sales in France had been slightly more than half the total for the group, but in 1975, for the first time, sales outside France were greater (52.5 per cent) and yet more than 62 per cent of the refining capacity was located in France under the control of CFR. Despite the fact that some of CFP's overseas markets could have been served from France (the distribution of overseas sales was approximately: Germany 15 per cent, Benelux 10 per cent, United Kingdom 20 per cent, Italy 25 per cent, Africa 14 per cent, United States 7 per cent, others 9 per cent)[13] over 90 per cent of CFR's sales took place in France[14] (Appendix 4). Against the imbalance in refining could be placed the fact that the French market, by deliberate government policy, had always been more profitable than other markets in Europe and three of CFP's important overseas markets were distinctly unprofitable. One of these was Italy where devaluation of the lira against the dollar and price controls on petroleum products meant that product prices were not aligned with new crude prices; this had made the market so unprofitable that CFP had threatened to withdraw. The two other unprofitable markets were the Netherlands and Germany where very fierce competition kept prices the lowest in Europe. CFP's profitable African operations were subject to restraints on the repatriation of capital and could not therefore compensate the downstream losses incurred in Europe outside France.

A further heavy drain on the company's financial resources was the development of the North Sea gasfield, Frigg. Production from Frigg was expected to start in late 1977, by which time development would have cost over FF15 billion[15] of which CFP would have supplied about FF4.5 billion. Even the expected initial production rate, which would subsequently increase rapidly, would bolster CFP's gas-production figures and its cash flow (see Appendix 5).

After representing around 22 per cent of total investment in the early seventies, in 1974, exploration and production represented 35 per cent of investment, at FF2.3 billion; 67 per cent 1975, and the same figure in 1976; the major part of this increase was represented by Frigg. Besides these investments CFP's other activities had taken on a lesser importance. Refinery investment had inevitably slowed down, and transportation investment, apart from the acquisition of CAN, was subject to great restraint for the same reasons.

In petrochemicals, which consistently through the seventies represented around 8 per cent of sales, one important investment was made in 1974. CFP, through CFR, acquired an 87 per cent share in Hutchinson–Mapa which was France's largest manufacturer of industrial rubber products, and which employed about 13,500 people in North Africa, Germany, Belgium, Spain, Italy, and Switzerland. Through its holding in SOCABU, CFP already had rubber products to sell and with the new acquisition CFR had an important position in the rubber market. The sales of the new acquisition (FF671 million in 1974) were equivalent to a quarter of ATO Chemie's sales in 1974, and a third in 1975 (a bad year for ATO). But the acquisitions of Hutchinson–Mapa and CAN were single investments. In 1976, due to the demands of exploration and development, every other investment budget except marketing was lower than in 1971 (see Appendix 6).

The stock market unsurprisingly showed its appreciation of CFP's cash difficulties by marking the share down strongly from the middle of 1973 on, until it was level with the official index, to which it had not been equal since 1964 (Appendix 7). In his address to the widely dispersed shareholders in June 1977 on the 1976 results, Grenier de Lilliac was understandably restrained about his company's immediate prospects. Such restraint by subsequent financial reports proved to be well justified through 1979.

Perspective

By 1980 the place of CFP among the ranks of the world oil companies was beginning to be defined with a little more distinction than could be made by a simple comparison of sales or assets figures, a process of distinction hastened by, if not initiated by, the events in the world of oil in 1973.

CFP was without a doubt a sizeable, fully integrated oil company. In historical perspective the way in which each sector of activity had been undertaken was so deliberate that an observer had even called the creation of the group a process of 'immaculate conception'.[16] It had of course been fortunate that oil was struck so early in the company's history and that CFP had for so long been associated with large and successful oil companies whose collective success at finding oil had been very much to the benefit of CFP. Although overwhelmingly dependent on the Middle East as a result of these relationships (see Appendix 2) CFP had nonetheless achieved exploration successes in its own right.

The first notable discovery had been in Algeria of both oil and gas, and the most recent had been the North Sea gasfield of Frigg. In the British North Sea an oilfield, Alwyn, was still being evaluated and in Indonesia an oilfield had been developed which was expected to contribute 1 million tons a year from 1976 onwards to CFP (about 1.5 per cent of total production). In North America, a promising exploration program in Labrador was progressing, and in the United States CFP had a number of small production operations.

The early and continued discovery of oil enabled CFP to expand and develop somewhat independently of the government. CFP had not been immune to government influence, but it had needed no help in surviving and was able to continue to attract private investment as well as generate enough earnings internally. The direct government presence, two commissioners on a board of eleven, was restrained. Although the board could make decisions with a three-quarters majority, the commissioners could suspend but not veto any decision, although they had never exercised this right. In this respect CFP was not unlike its British analog and partner in IPC, British Petroleum, although all business in France, including oil, was more responsive to government direction than in Britain.

Since the middle fifties CFP had been, and had become increasingly more so, long on crude. Initially this had pushed CFP to open new markets to absorb production but, by the seventies, the shortage of refining capacity with respect to crude supply was seen by one executive as a deliberate policy, although this was unlikely. What was more certain was that the comfortable position on crude supply had led the company to be one of the less impressive performers in the oil industry by financial criteria. Again in this it had much in common with BP. In comparisons between CFP and the European oil multinationals and independent American and European national oil companies of roughly similar size to CFP, it was noticeable that CFP's performance could at best be called 'lack-lustre'. This could have been a reflection of the absence of the need for really aggressive management to survive in the market. In France, the principal refining subsidiary CFR had a special position despite the controlled competition that existed. Although it had never achieved it, by 1977 CFR was close to supplying the 25 per cent of the French market to which it had a right (1976 share of refining 25.6 per cent, share of market 23 per cent).[17] Growth in CFR's distribution had been principally at the expense of the small French distributors who had effectively been squeezed out of the market by the government's oil policy. In contrast to the Elf group, the arrival of TOTAL on the French market was essentially an exercise in acquisition and not in aggressive growth of market share, and this may have left its imprint on CFP's management style. Outside France, CFP had no single large and profitable market and the close and beneficial relationship with the major partners in IPC may have lulled CFP into a false sense of security with its Middle East production, from which it had diversified very little.

With the proviso that operations in different markets make close comparisons subject to caution, a cross-company comparison (as shown in Appendix 8) gave indications as to why CFP had never achieved the stature in the stock market comparable to the multinationals, despite the fact that in operating terms it was a respected competitor. Its returns on investment and returns on sales were lower than its peers and competitors; its liquidity was lower and its gearing higher although in part this could be explained by the fact that it was European and was, in the seventies, overcommitted to exploration investment (Appendixes 9, 10, and 12).

What may have been strengths in the fifties and sixties had proved to be points of vulnerability in the post-1973 atmosphere. With the quasi-totality of CFP's supply coming from OPEC countries the nationalization of production took away a major source of profit and cash flow and the subsequent drop in consumption in the Western world left CFP greatly overprovided with shipping capacity. Whereas the combined tonnage of CNP and CAP in 1975 had gone up 65 per cent since 1971, tonnage carried had dropped back to the level of 1971. Some oil multinationals suffered in much the same way in these respects but they had other strengths not shared by CFP.

As mentioned above the downturn in consumption in turn adversely affected refining, with the greatest exposure (i.e. undercapacity) in France and Italy. These two markets were also in serious difficulties because, as a result of the energy crisis, the new energy policies of each government were designed to reduce oil consumption heavily, a change that would be felt most dramatically in fuel oil consumption for electricity generation. CFR was a large supplier of Electricité de France (EDF) and reduced consumption of fuel oil would require new investment in cracking facilities which would be difficult to finance in a period of undercapacity. In the German and Dutch markets particularly, CFP suffered another disadvantage in that it had no indigenous production and was competing against the oil multinationals who had both domestic and American production resources. The rise in the price of the principal energy source, Middle East oil, had served to provide those companies who had production facilities elsewhere with a source of windfall profit with which to subsidize competition in the market place. By virtue of having such an extensive foreign distribution network, and thus competing everywhere with the multinationals having the advantage of these windfall profits, CFP was especially vulnerable to the competition.

For many businesses, and the oil firms were no exception, the protection against downturns in one sector of activity could often be achieved by diversification, but CFP had diversified very little. The now-nationalized fields in Algeria apart, CFP did not produce gas in sizeable quantities, although the Frigg production would change this. Diversification into petrochemicals had been small, even half-hearted, and contributed less than 10 per cent of sales. Other nonoil activities included the purchase in 1975 of a coal mine in South Africa whose production of 3 million tons annually from 1979 would go wholly to Japan and Belgium, and the establishment in 1975 of a joint (50 per cent) subsidiary with Pechiney Ugine Kuhlman to develop nuclear energy interests via the exploiting of uranium deposits in France, South West Africa, Gabon, and Niger; the three African interests were already producing uranium oxide concentrates.

In 1980, in a bid to increase France's stake in mining operations, CFP purchased 8.5 per cent of COFRAMINES, the government's principal venture in domestic and foreign mining exploration and production. It is expected that COFRAMINES will concentrate its activities in Saudi Arabi and Africa.[18] Small interests in fertilizers, solar power, and experimental agriculture completed the activities, but none of these interests was large enough to figure separately in CFP's annual sales figures.

According to a company press release CFP's exploration investments in new energy resources are expected to be $315 million in 1980 compared to $160 million in 1979 while development expenditures in newly discovered energy resources will be expanded to $570 million in 1980 from $380 million in 1979.

Despite its basic strengths the outlook for the immediate future was uncertain due to these adverse changes in the company's environment. CFP certainly had felt that its situation was special enough to warrant its association with the Brussels Memorandum, presented to the Commission in 1976 by five European national oil companies – CFP, Elf, ENI, Veba, and Petrofina.[19] The Memorandum publicity brought to the notice of the Commission the perilous position of European refining, the difficult market conditions, and the problem of these companies to survive in the markets without a beneficial price control system, which would allow them to continue to explore in costly areas like the North Sea, with its benefits to European security of supply. No official reaction to the Memorandum had appeared by early 1977. In France, the government had itself taken action which would involve it more closely with the future of the company. In early 1977 the number of directly appointed government members of the board was raised from two to four; one of the new appointees, M. Guillaumat, Chief Executive Officer of SNEA, was already on the board in his own right. The move was seen as a confirmation of the government's interest in the company strategy and an expression of the need to be seen to be in control (for domestic political reasons) and – long delayed – as being appropriate to its shareholdings. On the part of CFP, it was considered as providing the government with a better understanding of the company's problems.

The irony of CFP's position in 1977 was that as the first and most successful oil company established by a European government to provide an alternative to the American and Anglo-Dutch oil corporations, the energy crisis showed that CFP was not yet strong enough to survive without some form of government support, support not yet defined but for which the need was becoming ever more apparent and which would have to extend beyond the help already given to the company's main subsidiary, CFR. Grenier de Lilliac indicated in the 1976 Annual Report where such assistance might be applied. After noting that the company intended to continue exploration activities he observed that 'the financial effort to be maintained in this sector may necessitate public assistance.'[20]

Poor financial results continued into 1979 due to the non-growth, and even decline, of oil consumption throughout Western Europe in general and France in particular. Because of CFP's surplus in refining and shipping, the company showed losses in both operations. CFP's losses were particularly large in Germany and Holland where competition produced downward pressures on product prices. The profit and loss statement for 1977 was so adverse that CFP's President speculated in the Annual Report 'whether (the company) can continue indefinitely to supply markets where there is no recognition of a legitimate return on investments.' (He did not specify as to what he meant by 'legitimate'.) The situation improved somewhat in 1978 as product prices firmed in Europe except in France and the United Kingdom, obviously two large markets for CFP. Nonetheless the firming of some markets was sufficient to produce a somewhat improved financial result in 1978 – and an increase in the price of CFP's shares. When in the wake of the Iranian embargo and other market forces crude prices increased by 60–65 per cent in 1979, CFP's profits soared as compared to 1978. In this respect CFP followed the general pattern of the oil industry though the higher profits represented essentially inventory appreciation. However, there was some continued improvement in the margins on refining and marketing throughout 1979.

Appendix 1. Operating Data, 1956–70[21]

Figures in thousands of metric tons

	1956	1957	1958	1959	1960	1961	1962	1963	1964	1965
CRUDE OIL										
Resources	11,363	9,401	12,887	15,059	19,090	22,059	25,538	30,947	34,044	37,773
Use within the group	6,987	6,373	9,229	10,236	10,188	14,089	15,039	18,209	20,951	23,273
Use outside the group	4,376	3,028	3,658	4,823	8,902	7,970	10,499	12,738	13,093	14,500
REFINING										
Crude oil processed	9,944	8,838	10,760	11,535	13,551	15,404	16,169	18,783	21,144	23,135
DISTRIBUTION										
Finished products marketed	8,620	7,657	9,824	10,677	13,363	15,131	16,654	19,399	21,623	24,026
MARINE TRANSPORTATION (deadweight)										
Fleet operated by the group	714				1,017					1,376
of which CNP's own ships	263				503					687

	1966	1967	1968	1969	1970
CRUDE OIL					
Resources	42,000	44,000	52,000	57,000	61,000
Use within group	26,000	29,000	33,000	38,000	39,000
Use outside group	16,000	15,000	15,000	19,000	22,000
REFINING					
Crude oil processed	24,000	26,500	31,000	36,000	37,500
DISTRIBUTION					
Finished products marketed	26,839	29,149	33,034	37,424	40,547
of which in France	14,739	15,860	17,673	20,371	21,497
MARINE TRANSPORTATION (deadweight)					
Fleet operated by the group					1,584
of which CNP's own ships					1,043

Appendix 2. Crude Oil Resources[22]

Figures in thousands of barrels per day

	1972	1973	1974	1975	1976	1976/1975 Variation (per cent)
Iran	260	288	307	233	238	+2.1
Iraq	416	411	528	404	260	−35.7
Qatar	61	54	44	22	43	+95.4
Abu Dhabi (ADPC/ADMA)	300	348	335	308	275	−10.7
Abu Dhabi (Abu Al Bu Koosh)	–	–	8	33	43	+30.3
Oman	27	29	25	33	30	−9.1
Dubai	40	57	61	65	80	+23.0
Middle East	1,104	1,187	1,308	1,098	969	−11.8
Algeria	141	138	132	142	207	+45.7
Tunisia	3	2	2	1	1	–
Indonesia	–	–	–	4	28	–
North America	5	6	6	6	8	+33.3
South America	–	–	–	–	0.05	–
North Sea	2	1	1	7	10	+42.8
Other countries	151	147	141	160	254	+58.7
Own resources	1,255	1,334	1,449	1,258	1,223	−2.8
Purchases	187	286	323	221	318	+44.1
TOTAL RESOURCES	1,442	1,620	1,772	1,479	1,541	+4.1
USE OF RESOURCES						
Within the group	964	1,176	1,228	1,035	1,133	+7.5
Outside the group	478	444	544	444	428	−3.6
In France	699	790	825	617	712	+15.3
In other countries	743	830	947	862	829	−3.8

Appendix 3. Crude Oil Refining[2 3]

Figures in thousands of barrels per day

	1972	1973	1974	1975	1976	1976/1975 Variation (per cent)
In refineries owned by the group or in which the group has an interest (crude processed for or sold by the group)						
France	585	731	729	590	625	+6.0
Outside France	275	287	377	355	387	+9.0
	860	1,018	1,106	945	1,012	+7.0
In outside refineries						
France	48	50	16	–	–	–
Outside France	56	75	48	54	46	–14.8
	104	152	64	54	46	–14.8
Overall France	633	781	745	590	625	+5.9
Overall outside France	331	362	425	409	433	+5.9
OVERALL	964	1,143	1,170	999	1,058	+5.9

Appendix 4. Sales of Finished Products[24]

Breakdown of sales by main product category

	1972	1973	1974	1975	1976	
Petrol	180	194	189	236	249	+5.5
Avgas and jet fuel	37	39	41	53	56	+5.6
Kerosene and gas oil	121	128	141	136	143	+5.1
Fuel oils and heating oils	583	725	718	615	633	+2.9
Lubricants	10	9	9	7	7	–
Other products	84	98	97	110	122	+10.9
OVERALL	1,015	1,193	1,195	1,157	1,210	+4.6
SALES AREA						
In France	519	630	627	553	565	+2.1
Outside France:						
Western Europe	336	411	395	431	464	+7.6
Africa	75	73	77	84	92	+9.5
North America	39	36	40	42	47	+11.9
Other	46	43	55	47	42	–11.6
OUTSIDE FRANCE	496	563	568	604	645	+6.7
OVERALL	1,015	1,193	1,195	1,157	1,210	+4.6

Appendix 5. Natural Gas Production and Cash Flow[25]

Volumes marketed on a joint basis: CFP's share (thousands of cubic feet per day)

	1972	1973	1974	1975	1976	
North America	4,573	5,159	7,618	10,193	18,206	+78.6
Australia	–	–	–	1,250	1,655	+32.4
Holland	9,550	9,376	9,173	9,869	11,669	+18.2
Total	14,123	14,335	16,791	21,312	31,530	+47.9

Estimated initial production from Frigg 1977 532,000
CFP share 20 per cent = ... 106,400

Cash flow (millions of francs)

1972	1,810.8
1973	2,642.5
1974	4,078.4
1975	2,177.4
1976	2,876.2

Appendix 6. Group Net Capital Investment[26]

Figures in millions of francs

	1972	1973	1974	1975	1976
NATURE OF OPERATION					
Exploration	257.5	289.4	545.8	536.1	780.6
Production	183.6	312.3	917.5	1,759.7	3,279.2
Mining and nuclear	1.1	22.5	15.9	18.2	128.7
Transportation	227.1	266.2	821.2	160.6	43.6
Refining	589.5	917.7	697.3	283.8	244.5
Petrochemicals	240.5	101.4	753.9	239.3	287.8
Marketing	673.7	481.8	490.0	284.0	382.4
Real estate, commercial, financial, and other	0.1	94.6	−100.8	138.1	27.6
TOTAL	2,173.1	2,485.9	4,140.8	3,419.8	5,174.4
GEOGRAPHICAL AREA					
France	1,104.3	1,139.0	2,109.6	919.6	769.6
Other European countries	525.8	788.4	1,085.6	1,511.5	2,140.8
North Africa	23.3	77.3	201.6	145.1	743.6
Other African countries	93.3	129.6	60.6	42.7	189.2
Middle East	170.0	153.8	267.0	433.9	466.1
America	63.5	102.4	223.1	256.0	539.2
Asia–Australia	192.9	95.4	193.3	111.0	325.9
TOTAL	2,173.1	2,485.9	4,140.8	3,419.8	5,174.4

Appendix 7. Share Price Movement[27] and Capital Structure[28]

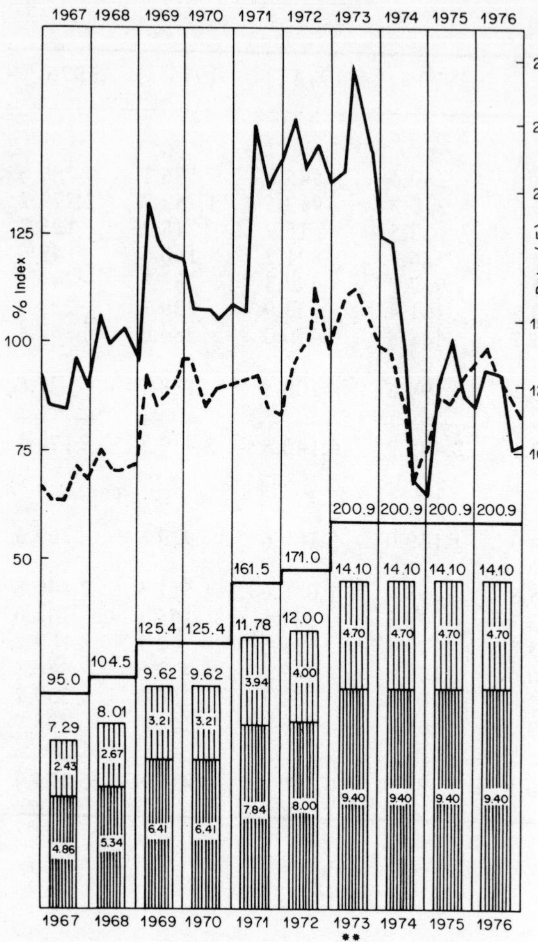

CFP capital structure is as follows:	(per cent)
Desmarais Frères	1.93
Banque de Paris et des Pays-Bas	1.37
Société Française des Pétroles/BP	1.00
Banque de l'Union Parisienne	1.20
Compagnie Auxiliaire de Navigation	0.98
Compagnie d'Assurances Générales	0.60
Crédit Lyonnais	0.56
Lille Bonnières-Colombes	0.52
French state	35.00
Others	56.84

Appendix 8. Company Comparisons, 1975[29]

	RD/Shell	BP	CFP	SO Indiana	Continental	Elf	Phillips	Petrofina
Sales and other revenue ($ millions)	29,154	15,718	8,732	9,955	7,253	6,841	5,134	4,037
New investment incl. exploration expenses ($ millions)	3,480	1,594	762	1,876	797	1,868	694	372
Employment (in thousands)	161	78	44	47	44	23.4	30.5	20.8
New investment as a percentage of total revenue	11.9	10.1	8.7	18.8	11.0	27.3	13.5	9.2
Total revenue per employee ($)	1,813	2,015	1,984	2,118	1,648	2,924	1,683	1,941

Appendix 9. Net Profitability[30]

(Net income[a]/shareholders' equity[b]) x 100

	1965	1970	1975
Royal Dutch/Shell	8.6	9.5	18.5
British Petroleum	10.2	8.9	6.4
CFP[c]–Group TOTAL	9.0	10.3	9.7
SO Indiana	8.4	9.8	15.5
Continental	10.2	11.2	17.4
Elf-Aquitaine	–	1.7	13.9
Phillips	10.4	7.1	17.5
Petrofina	13.5	15.4	21.1

[a]Net income (excluding minority interests).
[b]Shareholders' equity before allocation of income.
[c]CFP, although the commonly used name for the group, is used by the group to refer to the parent holding company, whose results are not used here. Figures are for the consolidated TOTAL group.

Appendix 10. Return on Sales[31]

(Net income/total sales) x 100

	1965	1970	1975
Royal Dutch/Shell	10.1	9.4	7.1
British Petroleum	9.8	6.8	2.2
CFP–Group TOTAL	6.4	5.4	1.8
SO Indiana	9.0	8.6	7.9
Continental	6.8	6.3	5.0
Elf-Aquitaine	–	2.7	4.9
Phillips	8.8	5.9	6.7
Petrofina	5.4	4.9	4.2

Appendix 11. Debt Ratio[32]

(Long-term debt/total current liabilities) x 100

	1965	1970	1975
Royal Dutch/Shell	9.6	13.4	21.0
British Petroleum	15.0	23.0	18.0
CFP–Group TOTAL	24.0	32.0	41.0
SO Indiana	11.0	19.0	22.0
Continental	25.0	28.0	25.0
Elf-Aquitaine	NA[a]	25.0	38.0
Phillips	18.7	27.2	24.4
Petrofina	45.0	46.0	52.0

[a]Not available.

Appendix 12. General Liquidity Ratio[33]

Current assets/current liabilities

	1965	1970	1975
Royal Dutch/Shell	2.03	1.56	1.61
British Petroleum	1.51	1.29	1.31
CFP–Group TOTAL	1.29	1.23	1.21
SO Indiana	2.33	1.53	1.47
Continental	1.80	1.66	1.39
Elf-Aquitaine	NA[a]	1.54	1.19
Phillips	2.02	1.54	1.80
Petrofina	0.85	0.83	1.22

[a]Not available.

References

1. 'The Fifty Largest Industrial Companies in the World', *Fortune*, August 1976.
2. Henry Bérenger, *Le Pétrole et la France*, Paris, 1920, p. 176.
3. Richard F. Knisel, *Ernest Mercier*, University of California Press, 1967, pp. 33–44.
4. Richard F. Knisel, *Ernest Mercier*, University of California Press, 1967, p. 31.
5. Private communication to author, 1977.
6. *The Oil Era, TOTAL in the Oil Industry*, CFP publication, 1966, p. 15.
7. *La Raffineur Français*, CFR Publication, 1977, p. 34.
8. *The Oil Era, TOTAL in the Oil Industry*, CFP publication, 1966, p. 15.
9. *The Oil Era, TOTAL in the Oil Industry*, CFP publication, 1966, p. 15.
10. Compagnie Française des Pétroles, Exercise 1970.
11. Compagnie Française des Pétroles, Exercise 1974.
12. Compagnie Française des Pétroles, Exercises 1971–75 inclusive.
13. Private communication to author, 1977.
14. *Le Raffineur Français*, CFR publication, 1977, p. 24.
15. Private communication to author, 1977.
16. Jean Goudot, 'European Multinationals', *Business History Review*, Autumn 1974, p. 287. See also Lawrence G. Franko, *The European Multinationals*, Harper and Row, London, 1976, pp. 26–27.
17. Compagnie Française des Pétroles, *Annual Report*, 1976.
18. *Wall Street Journal*, March 10, 1980.
19. 'Oil Firms Seen Challenging Seven Sisters', *International Herald Tribune*, September 24, 1976.
20. Compagnie Française des Pétroles, *Annual Report*, 1976.
21. *The Oil Era, TOTAL in the Oil Industry*, CFP publication, 1966, p. 41 and Exercise 1970.
22. Compagnie Française des Pétroles, *Annual Report*, 1976.
23. Compagnie Française des Pétroles, *Annual Report*, 1976.
24. Compagnie Française des Pétroles, *Annual Report*, 1976.
25. Compagnie Française des Pétroles, *Annual Report*, 1976.
26. Compagnie Française des Pétroles, *Annual Report*, 1976.
27. Compagnie Française des Pétroles, *Annual Report*, 1976.
28. Private communication to author, 1976.
29. Company Annual Reports.
30. Private communication to author, 1976.
31. Private communication to author, 1976.
32. Private communication to author, 1976.
33. Private communication to author, 1976.

CHAPTER 4

Société Nationale Elf-Aquitaine

In September 1976, a significant change took place in the French petroleum world. The wholly publicly owned Etablissements de Recherches et d'Activités Pétroliéres (ERAP) restructured its holdings in the Elf–ERAP group of companies and in the mixed private and public company Société Nationale des Pétroles d'Aquitaine (SNPA) to create a new group, Société Nationale Elf-Aquitaine (SNEA), in which ERAP's share of the capital was 70 per cent. ERAP became a pure holding company and SNEA took over the assets of both groups as an operating company. The new company was the thirty-fifth largest in the world and the third largest in France, with sales in 1975 of FF30.7 billion and was exceeded in size only by Renault (1975 sales FF33.5 billion) and Compagnie Française des Pétroles (1975 sales FF39 billion).[1] SNEA's interests ranged across most of the world and covered gas and sulfur production, oil exploration, production, transport, refining and marketing, petrochemicals, pharmaceuticals and non-ferrous mining, particularly nickel.

The arrival of SNEA among the ranks of the world's largest companies was the result of growth in different ways of the two merging groups, Elf–ERAP and SNPA. For both companies, their evolution was intimately linked with, and was an outgrowth of, French oil policy (which is treated separately in Chapter 2). This interventionary and gradualist policy of the French government aimed at creating a powerful national oil company which could work on an international scale and respond to the competition offered by the large multinational oil companies, and at the same time have French national objectives as one of its principal frames of reference.

The very size of SNEA made it an important part of French industry, but in the new environment created by the 1973 oil crisis, issues had risen which the French oil policy alone was not equipped to deal with. Despite its solid quality the new group could expect the immediate future to bring problems of a scale not encountered in the successful growth period, and which could severely test the group's apparent strength.

The Search for French Oil

Before World War II, in addition to the work by Compagnie Française des Pétroles (CFP) in the Middle East, various initiatives were taken by the French government in encouraging oil exploration on French territory. Encouragement to form exploration companies was often given directly, as in the case of colonial administrations, notably Morocco, or indirectly by a government-financed policy and administrative body, Organization Nationale des Combustibles Liquides (ONCL), which encouraged exploration in France as well as overseas. In 1937, under the auspices of this body, an exploration company Centre de Recherche de Pétrole du Midi, started work in Southwest France and in July 1939 its efforts were rewarded when gas was struck at Saint Marcet. To exploit the find and explore further, a new company was set up in the same month by decree, Régie Autonome des Pétroles (RAP).[2] RAP, publicly owned, was an innovation by the state, as it was the first state-owned company to have a specific and direct industrial and commercial role (as opposed to an indirect and exploratory role) in the field of hydrocarbons.

The outbreak of World War II in September 1939 not unsurprisingly slowed down exploration efforts, and RAP, despite receiving a concession for further exploration around Saint Marcet, made no additional discoveries. The Vichy government gave an adjoining concession to a new company, Société Nationale des Pétroles d'Aquitaine (SNPA), part public, part private, but it too made no discovery before 1945. At the end of the war General de Gaulle, on his return to power, considered that no company yet existed which wholly responded to the state's requirements of control of a vital resource, and the only significant French company that existed, CFP, had enough problems recovering from war damage. It was also part privately owned, which might have made it an unsuitable vehicle for national objectives.

Thus General de Gaulle in 1945 founded a new body, Bureau de Recherches des Pétroles (BRP). An 'établissement public' (publicly owned company) like RAP, BRP was charged with animating an exploration program on French territory by encouraging exploration companies with equity and policy development. Unlike RAP, it could only undertake commercial activity indirectly, by investing in commercial ventures. It took over responsibility for RAP and controlled the states' 51 per cent holding in SNPA. The man who was appointed head of BRP as president had been specifically concerned with oil matters as a Directeur des Carburants in the Ministry of Industry. He was Pierre Guillaumat, who was then thirty-six years old and whose relationship with de Gaulle was close. They were personal friends and de Gaulle had served under Guillaumat's father in the army. BRP was to look for oil 'in the exclusive interest of the nation'.[3]

An intensive drilling program started at home and overseas. BRP, funded by grants from the state, which by 1950 amounted to FF560 million,[4] invested in a number of operating companies, and its efforts were augmented by similar investments made by RAP, which had been receiving an income from the Saint Marcet gasfield since 1943, and by SNPA, which too had been receiving a sales income following a small oil strike at Lacq in 1949, also in Southwest France. The most important investments proved to be in Algeria and Equatorial Africa. In the latter area, BRP acted through Société des Pétroles d'Afrique Equatoriale (SPAFE) in which the other investors were private banks and financial institutions. Exploration was carried out in consortia with Mobil and Shell. In Algeria, BRP and RAP both made investments in conjunction with the oil companies, most notably Shell of the majors and various American and European independents. As the principal animator of the French exploration program, BRP took an essentially 'open door' approach and aided and cooperated with any interested and competent oil company. The most important joint companies formed proved to be Compagnie de Recherche et d'Exploitation du Petrole du Sahara (CREPS – 65 per cent RAP, 35 per cent Shell) and Société Nationale de Recherche et d'Exploitation des Pétroles en Algérie (SNRepal), which combined the interests of BRP, CFP, and the Algerian colonial government, but there were other companies in which BRP, RAP, and SNPA had interests (see Appendix 1).

By 1950, with the exception of the small oil find by SNPA, no results had been obtained from BRP's efforts and in 1951 the president, Pierre Guillaumat, left to become Director General of the Atomic Energy Commission. However, exploration activity did not slacken and began to bring results. In 1951, SNPA found a large gasfield in Aquitaine; in 1954, an Esso company in which BRP had a 10 per cent share (ESSOREP) found oil in Les Landes; and in 1956 very big finds of oil and gas were made in Algeria and of oil in Gabon. Other discoveries followed, in the Congo (1957) plus some small finds in the Paris region, but the Algerian discoveries were the most important to BRP. The faith of government and private investors seemed justified and repayments of the amounts devoted to exploration now seemed to be a reality. Since 1953, the government had added to the grants to BRP for its investments part of the proceeds of a sales tax on petroleum products ('Fonds de Soutien'), and this was to provide a regular income to BRP for some years ahead until it was dwarfed by receipts from sales of oil and gas.

Sales were, however, overshadowed by events in Algeria. In 1958, General de Gaulle had been returned to power specifically to deal with the Algerian war, and Guillaumat had left the Atomic Energy Commission to become Minister of the Army. In the next two years, General de Gaulle led the French army to military victory over the Algerian rebels and in his subsequent political *volte face* in opening negotiations for independence with the Algerians (and the subsequent granting of it in 1962), oil matters became a central concern. From 1960 Guillaumat was again involved with them.

UGP – ERAP – Elf

From the time of the discoveries in Algeria, Gabon, and the Congo, the preoccupation of BRP had been with disposing of the oil. The price of the oil was more expensive than Middle East oil, and there was a buyers' market. None of the BRP companies which had found oil had refining or marketing operations in France or French territories and the marketers already established on the French market (including CFP) were reluctant to take the more expensive supplies.

As part of a French government program to assist the sale of Franc Zone crude, in 1960 BRP changed its role from an animator of exploration companies to a holding company for producing companies. At the same time, a new company was created, Union Générale des Pétroles (UGP), the shareholders of which were RAP, SNRepal, and Groupement des Exploitants Pétroliers – BRP's various subsidiaries including SNPA. Its task was to refine and market the production of the French companies who were frustrated, as noted above, in bringing their new oil to the markets. Its president was Guillaumat and it was, like BRP, an 'établissement public'. In addition, Guillaumat reassumed control of BRP and his brief, in conjunction with running UGP, was to rationalize the companies in the petroleum sector in which the government had interests through BRP, RAP, and SNPA.

The government did, in fact, have the power under its oil regulations to impose the Algerian crude on French marketers ('devoir national') and the creation of UGP was not, in terms of disposing of Franc Zone crude, absolutely necessary. The foreign private companies operating in France (principally Esso, Shell, BP, and Mobil) took the news as a clear indication that the government intended to create a national oil company and that the government intended to extend its intervention in the market beyond assistance to exploration and general overall control of the internal market. They, like CFP, were apprehensive about the *dirigiste* methods that would be used to foster the growth of UGP.

Guillaumat's return to the center of the stage for oil affairs (although he retained his ministerial functions) was welcomed by many, who felt he had the drive to create a truly successful national oil company, and he moved quickly. UGP bought both market share (4 per cent) and refining capacity by purchasing the French operations of Caltex (which was owned 50 per cent by Texaco and 50 per cent by Standard Oil of California) the same year as UGP's creation. A subsidiary separate company, Union Industrielle de Pétrole (UIP) operated the refinery at Ambès, Bordeaux, on behalf of the two owners, UGP (60 per cent) and Caltex (40 per cent). It was the start of a long amicable relationship between the two American multinationals and the state company.

In 1962, Algeria was granted independence, and a central issue of the Evian agreements with Algeria in March of that year was the role of the French oil companies in the country. Under the terms of the agreement French companies were given preferential treatment and BRP and its sister companies were the most direct beneficiaries, accounting then for 65 per cent of Algerian production (81 per cent including the production shared with Shell – see Appendix 1).

The apprehension of the private companies to the implications of the creation of UGP was realized in 1963 when the new round of import quotas was announced by the government. 'Devoir national' obligations were imposed on all French marketers to buy Franc Zone crude, almost entirely available from BRP and associated companies and at a price which was considered to contain an element of indirect subsidy to the state companies. The obligation could be offset by the amount of Franc Zone crude each company produced, which was of some advantage to Shell and CFP having Algerian production and Esso with production in France. Approximately three-quarters of the volume available was disposed of under this obligation and the remainder was to be refined and marketed by UGP. The anxiety of the other companies over the behavior of UGP was compounded when the company began to look aggressively for market share. Previously there had been an industry agreement between all refiners relating to market position, refining capacity and import authorization, the effect of which was to make market expansion possible only by merger, not by competition. The competition given by UGP, not party to the agreement, caused Esso to take the initiative in 1964 in breaking up the arrangement. Thenceforth competition was more intense in the French market.[5]

UGP having been established, attention was then directed to the multiplicity of companies controlled by the state in various overlapping and interlinking areas of the oil industry. The government wished to rationalize the administrative structures under one titular head, and in 1965 a major act of consolidation was achieved when BRP, RAP, SNPA, and UGP were brought together under the control of one 'établissement public', Entreprise de Recherches et d'Activités Pétrolières (ERAP) with Guillaumat as president. Although it was also to act as the vehicle for Franco-Algerian cooperation, it acted principally as a holding company for the constituent companies, but with an active brief to streamline and coordinate the overlapping activities. (See Table 1.)

SNPA, in view of its size and its private shareholders, continued to be run as a separate entity. (The development of SNPA is dealt with in greater detail later. Reference to ERAP in the text includes SNPA unless otherwise noted.) Most of the other companies (and their many subsidiaries) disappeared under the name of ERAP. An observer at the time said 'Mattei made ENI and looked for oil; BRP found oil and Guillaumat made ERAP',[6] but this was neither fair to the Italian company nor BRP. ENI had had the advantage of profitable protected domestic gas production for its early growth and BRP, while having interests in both RAP and SNPA which were similarly advantaged (particularly the latter), had grown under consistent political encouragement from whatever had been the French regime in power.

The new group was still receiving financial support from the government.

Table 1. Original constituent companies of ERAP

Company	Subsidiary	Activity
BRP	CREPS, SPAFE Gabon, SOFRepal SPAFE Congo, SAFRAP, SNRepal	Oil exploration and production
RAP		Oil and gas production
SNPA	SNGSO, CeFeM, ORGANICO	Oil and gas exploration; oil, gas, and sulfur production; gas distribution, petrochemical production.
UGP	UIP, Cie Nat. de Navigation	Oil refining and distribution, and transportation.

Receipts from the petroleum sales tax had risen from FF20 million in 1960 to FF220 million at the end of 1965.[7] These were sufficient to cover approximately 22 per cent of capital requirements of all the companies, SNPA included, which itself, by virtue of its healthy cash flow from gas production, supplied half of the residue and was self-financing (see Appendixes 2 and 3). SNPA had received income from oil sales since 1949, and from gas sales since 1957; the latter by 1966 was nearly forty times the income of RAP from similar sales, which were declining,[8] see Appendix 4). Income from oil for the other BRP companies had also begun in 1957, but large volumes only began with the decade of the sixties (see Appendix 1) when oil and gas sales together began to transform the balance sheet of the group. (See Table 2.)

Notwithstanding this, from 1966 when ERAP was created, the state increased its support to the group through the 'Fonds de Soutien' in order to encourage the investment that was necessary to get UGP on its feet and to diversify oil supply away from a large dependence on Algeria. As before, rising sales increased internal financing and the importance of the direct state support diminished (see Appendix 2). The increased funds were also to help ERAP undertake what was then the relatively innovatory task of providing service contracts to producer countries for exploration and production in exchange for long-term crude contracts; Iraq was the first country to benefit from this arrangement, which continues to the present day and which was instrumental in achieving an amicable agreement between France and Iraq when the assets of CFP in the country were nationalized in 1972.[10] Activities were increased in Equatorial Africa, Tunisia, and the North Sea; SNPA started prospecting in North America, and between 1965 and 1970 the group spent about 10 per cent of total worldwide investment on exploration expenses.[11] Large sums were also spent developing the Algerian fields and the pipelines serving them. The fleet was built up to 1.2 million DWT by 1969, mostly in the name of Compagnie Nationale de Navigation, a company in which ERAP had a minority share; refinery capacity grew from the original 1.5 million tons per year purchased from Caltex to 20.8 million tons per year by the end of 1969,[12] with refineries at Bordeaux, Lyon, and Paris as well as six overseas territories (see Appendix 5).

The diversity of interests and activities for which ERAP was responsible and the brief of ERAP for rationalization of the companies made a trademark essential, particularly at the distribution outlets where seven different trademarks existed.[13] In April 1967, fifteen months after ERAP was created, the mark Elf appeared and thenceforth the group was known variously as ERAP, Elf–ERAP, or simply Elf. UGP became Elf Union and, while UIP remained unchanged, the

Table 2. ERAP group oil production totals[9] (millions of tons)

1961	1962	1963	1964	1965	1966	1967	1968	1969
8.8	11.3	13.7	15.0	15.7	17.6	19.5	20.6	21.7

distribution arm in France became Elf Distribution. SNPA retained its distinct identity as 'Aquitaine' for its various activities. The new trademark brought homogeneity to the group, particularly for marketing, and this was needed not only because of the large number of original companies but also because some growth by merger had taken place; Elf–ERAP had absorbed a number of small independent distributors. Market share had also been increased with the assistance given by the oil policy of the government which favored Elf–ERAP in matters such as import quotas and the siting of service stations.

In 1968, ERAP took over Société des Huiles Renault, and brought the market share of the group for petroleum products up to 12.5 per cent.[14] A more significant move took place in 1970 when the biggest of the French independents, SOCANTAR, the holding company with the operating arm, ANTAR, sold 95 per cent of its capital. ERAP took a 41 per cent interest, the state a 10 per cent interest, and the other shares were taken by CFP, 24 per cent, and Caltex, 20 per cent. This immediately increased the share of the petroleum products market from 14 to 23 per cent and the share of the lubricants market from 8 to 18 per cent (see Appendix 6), with sales generally in geographic areas (North and East France) not covered by Elf. (See Table 3.)

From the outset ANTAR's operating results were consolidated with those of ERAP, but for management and marketing purposes ANTAR remained a distinct entity, in the same manner as SNPA. This was in part due to the fact that ANTAR had a number of long-term contracts with suppliers, notably CFP, and immediate absorbtion of the group was not possible. Supply to the new joint refining capacity of 30 million tons from the group's own sources in 1970 amounted to 23.6 million tons of oil (including 2.6 million tons by SNPA) despite interruptions in Nigeria due to the civil war (where Elf had started operations in 1960). Of this amount 16.5 million tons (80 per cent) came from Algeria and Algerian oil comprised 27 per cent of all French imports.

The nationalization with compensation of 51 per cent of French companies' oil assets and 100 per cent gas and transportation assets by the Algerians in February 1971 was unsurprisingly a hard blow to the group. Not only was Algeria a major oil source for ERAP but ever since independence it had been a source at a discount per barrel for French companies, although this had been reduced by the Algerians through 1969 and 1970. The nationalization could not have been wholly unexpected, for the assets of Shell, the second biggest single

Table 3. ANTAR and Elf, 1970[15]

	ANTAR	Elf	
Refineries	3	4	
Refining capacity	10	20	million tons
Outlets	6,120	5,570	
Fleet	0.6	1.13	million DWT
Sales	1.4	6.6	million francs

operator in Algeria, had been nationalized in 1970. Nonetheless the first reaction of Guillaumat was that he was only interested in full compensation for what he saw as full nationalization.[16] The initial negotiations were carried out between governments at Algeria's request, but by April 1971, the French government declared itself impotent and handed negotiations to the companies. The task of CFP in coming to agreement with Sonatrach, the Algerian national company, was simpler than that of Elf–ERAP because of its less complex company holdings and agreement was reached in July 1971. Elf–ERAP, with more at stake, concluded its agreement in October of the same year, but not before strong personal tensions had required a change in negotiators.

It was to be expected that the immediate reaction to the change in the fortunes of the group was sober. Some 600 people left the service of the group[17] and outside observers speculated on its future viability. The diversification of resources in the previous decade, particularly in the second half, augured well and in an internal report the management suggested that with various changes, particularly the bringing together of the SNPA and ERAP exploration teams, the group would remain viable.[18] Guillaumat felt that by 1975 oil production would re-attain 1970 levels of 480,000 barrels per day.[19] On July 29, 1971, the French Cabinet under M. Pompidou's leadership reaffirmed the *raison d'être* of ERAP's existence and confirmed its continued assistance by the state for the special tasks assigned to it in its role as a national company[20] and as a vehicle for bilateral relations with producer countries. It also confirmed the desirability of having only two French groups, CFP and ERAP, and gave its blessing to the full incorporation over time of ERAP's most important and independent subsidiary, SNPA.

SNPA

Unlike its principal shareholder Elf–ERAP (51 per cent), Société Nationale des Pétroles d'Aquitaine (SNPA) was not an evolved grouping of companies with separate but interrelated activities; SNPA, while part of the ERAP group, was an integrated whole able to stand on its own.

The company had been started by the Vichy government in 1941, to supplement the efforts of RAP in Southwest France in looking for hydrocarbons, together with another company, Société des Pétroles de la Garonne; but little or no activity took place in the war years. In bringing together private and public interests in almost equal proportions (49 per cent and 51 per cent respectively), SNPA was representative of a strand of French oil policy which wanted to associate private investment with the public interest in developing a national resource. When BRP was created, the control of SNPA was placed in its hands, and the two organizations shared a number of common board members. However, SNPA was managed as an independent entity as befitted its private investors.

After the war, exploration began in earnest and in 1949 SNPA found a small oilfield (3.5 million tons estimated reserves) at Lacq in Southwest France. While only expected at best to produce about 3 per cent of the French consumption (in

1950, 9.2 million tons per annum),[21] the find started producing in 1950 and it both financed and encouraged further drilling. In the same area, but seven times deeper, a large field of heavily sulfurated gas was found in 1951. 180,000 million cubic meters of gas and 50 million tons of sulfur were calculated to be recoverable over twenty-five years. SNPA was already the part owner with Gas de France and RAP (all holding 33 per cent of the shares) of a gas distribution company, Société Nationale des Gas du Sud-Ouest (SNGSO), which took the gas from the RAP field. The size of the new find was such that another company was created, Compagnie Française du Méthane (CeFeM), jointly owned by Gas de France (GDF) and SNPA, which was to take production to East and Central France, and Paris, with SNGSO just serving the Southwest. In the period immediately following the war, companies in which the state's interest was more than 50 per cent had been exempted from nationalization and this had been done to allow companies like RAP access to the markets, and *ipso facto* to profitable operations. SNPA, via SNGSO and CeFeM, was to benefit similarly. With the infrastructure and the treatment plant for sulfur extraction completed, by 1957 SNPA was able to start gas delivery, and soon through the two companies, was supplying nearly 100 per cent of French natural gas consumption, although the proportion fell through the years as consumption and imports increased (see Appendix 4). The sulfur production rose rapidly too and in the early years was an important revenue earner. The cash flow from sales of both products transformed the company's finances and immediately made it more secure for its activities.

In common with BRP, initially SNPA's primary interests were exploration and production. The early success at Lacq enabled the company to continue active development of these activities on a large scale, most notably in France and Algeria, where, following a discovery in 1958, SNPA became a modest producer of oil (see Appendix 1). By the early sixties SNPA was already considering diversifying away from oil and gas exploration and production, and had started investment studies for petrochemicals, based on the range of products including sulfur, which were extracted from the gas at Lacq.

The creation of UGP in 1960 gave for SNPA, as it did for BRP and RAP, an outlet for its Algerian oil production. As one of the Groupement des Exploitants Pétroliers, SNPA held 20 per cent of shares of the new company. For the next ten years, SNPA expanded steadily. Oil production in Algeria climbed slowly, and from 1966 was augmented by production in Canada, where SNPA had started operations in 1963. Gas and sulfur, the staple products, increased slightly in volume production at Lacq, but through the sixties were slowly displaced as the main revenue earners by growing oil production and by the petrochemical activities started in the early sixties. The result of these developments was that by 1970 the company was better integrated than at the beginning of the decade. Oil production was carried out in Tunisia (a new field), Algeria, and Canada, and gas production was steady at Lacq and being augmented by a large new nearby field found in 1968. Gas was also being produced and sold in Canada from 1970. Exploration investment had switched from being largely devoted to France at the beginning of the decade (79 per cent in 1962)[22] to a more equal distribution across the world, and investment in petrochemicals was yielding nearly a third of the companies' sales (see Appendix 7). The corner-stone of the petrochemical activities was a company set up jointly by UGP and SNPA which subsequently became known as Union Chimique Elf-Aquitaine (UCEA), of which SNPA owned 88 per cent. In 1969 operations had been extended into the organic field with a joint investment with CFP in order to form a company called Aquitaine Total Organico (ATO).

The nationalization of assets in Algeria in 1971 caused the oil production of SNPA to drop 30 per cent. With oil only accounting for 22 per cent of sales in 1970 (see Appendix 7), the diversified activities and the healthy cash flow (equal to and sometimes exceeding its investment requirements) enabled SNPA to absorb the loss. Of the companies controlled by ERAP, SNPA was certainly the strongest, which was very largely due to its profitable gas production at Lacq and perhaps also due to the discipline that the presence of private shareholders exerted. The advantages for bringing SNPA more closely into coordination with Elf–ERAP, with better usage of human and financial resources, were more than apparent, and the announcement by the Cabinet in July 1971 explicitly heralded the day when 'Elf–ERAP Aquitaine' would be an integrated whole.

The Years before the Merger

The immediate aftermath of Algerian nationalization was inevitable for the group as a whole and oil production plummeted to nearly half the previous year's volume; the balance had to be bought on the open market. This did little to improve the cash flow. Investment, however, hardly faltered. Further development took place on existing finds in the Congo, Nigeria, Tunisia, and North America. In the North Sea the Ekofisk discovery, in which the group had an 8 per cent interest, was being actively developed, and in 1971 further north a gasfield, Frigg, at least the size of Lacq had been discovered. The exploration teams of both Elf and SNPA were being drawn together into one. Investment was also taking place in expanding refinery capacity which grew from 32.3 million tons in 1970 to 39.4 million tons by 1972, and to 44 million tons by 1975. The fleet too was expanding, and from a size of 1.7 million DWT in 1970 was to reach nearly five times that size in 1976 at 5.4 million DWT.[23]

The slow recovery of the group was at first only slightly upset by the 1973 energy crisis. ERAP did not suffer nationalization of its major sources of crude, and as it was just marginally short on crude (1973, 17.1 million tons produced, 21.9 million tons sold) only a small part of its supply had to be bought on the open market. However, price control in France did not enable the group to pass on all its increased crude costs. This offset the increased revenue from Lacq, although under the government's price-freedom surveillance policy, gas prices were allowed to realign themselves slowly with those of oil. Demand for gas was still rising in France and the prospect of diminishing production from Lacq from 1983 coupled with a decreasing ability to supply the French market (SNPA's contribution was down to 40 per cent in 1975) had led GDF, as monopoly gas importers, to arrange for supplies from Algeria to be fed into the CeFeM system from 1965 onwards. The profits to SNPA from sales of gas were nonetheless considerable.

1973 was the year in which SNPA took steps to diversify further the thriving petrochemical activities. UCEA and ATO had already diversified into related products ranging from ethylene to high polymer plastics, and in 1973 SNPA created a holding and managing company, l'Omnium Financière Aquitaine pour l'Hygiène et la Santé (Sanofi), to control various interests, all of which were undertaken that year principally in the field of pharmaceuticals. They were Labaz SA, a pharmaceutical, hospital hygiene, and veterinary products manufacturer employing 3,200 people in and outside France, Laboratoire Michel Robilliard, a research group in immunology, and Yves Rocher, a cosmetic group. Sanofi also controlled a small group dealing with air conditioning and building insulation. All these companies were picked in part for their high growth, their low capital intensity, and their strong research interest,[24] but some were also part of a program of enlarging the industrial base and the employment around the Lacq gasfields. The town of Pau had tripled in size since the discovery of Lacq[25] and the gasfields, but attendant employment was expected to decline from the middle eighties.

SNPA had also for some years been diversifying its exploration expertise to look for minerals as well as oil, and in 1974 an important lead and zinc discovery was made in Australia. More significantly SNPA took a 50 per cent interest in Société le Nickel, holder of a large nickel concession in New Caledonia and the second largest nickel producer in the world. The reason for this investment was not clear. Le Nickel, an investment by Banque Rothschild, was losing large sums of money in 1972 and 1973, due to a slump in nickel prices, the devaluation of the dollar, and onerous tax conditions imposed by New Caledonia. In order to save the company and prevent it being sold into foreign hands, it is likely that the French government inflated SNPA's interest in mineral diversification, with a view to assuring long-term supply of a strategic mineral. Consequently in 1974, SNPA bought 50 per cent of the company with FF450 million, thereby retaining French control.[26] In 1975, with profitability still not in sight, SNPA had placed a moratorium on future investment in the mine, but production continued. However, mineral diversification was taken further and that year SNPA acquired a coal mine in Pennsylvania as well as pushing forward exploration in France, Australia, Canada, and South Africa.

Exploration was in fact a strength of the group as a whole, benefiting not least from the fact that exploration had been the first activity of SNPA and Elf–ERAP to be coordinated into one team (from 1972[27]). Several successes had been achieved in the North Sea, of which the most notable was the Frigg gasfield. In the Gulf of Guinea, where the group had had interests almost from its inception, continued discoveries in Gabon, the Congo, Cameroons, and Nigeria fuelled development of one of the companies' most profitable areas. In North Africa, Elf-Aquitaine was still exploring in Algeria and successful discoveries continued to be made in Tunisia and Libya. In North America the two Canadian offshoots of Elf and Aquitaine had each (particularly Aquitaine) been successful, and this seemed likely to continue. Unlike most of the other areas where the companies had been exploring, most work here had been done on land. Exploration also continued in France, encouraged in various ways by the government.

The pattern of production was very similar to that of exploration and the Gulf of Guinea, and in particular Gabon, was the most productive area. Crude production had improved greatly as Gabon and Tunisia came on stream, and in 1975 the group had been in sight of achieving the prediction at the time of the Algerian nationalization, of returning to the 1970 production figure in five years, only for production to fall back in 1976 as an agreement with Algeria was not renewed. This set back the date when the group would be self-sufficient in oil. In 1975 the group was 6 per cent short (23.3 million tons produced, 24.9 million tons sold) but in 1976 this was to increase to 35 per cent (18.2 million tons produced, 27.8 million tons sold).[29] In relation to their respective ownership of Elf-France (ERAP 80 per cent, SNPA 20 per cent) and their corresponding obligations to supply the refineries, SNPA was shorter on crude supply than Elf–ERAP (SNPA's own supply was only 15 per cent of the total – see Appendix 8) but whether this was of any significance it was difficult to say. In fact much of the production was sold profitably to premium markets other than France, and the bulk of supplies came to France from the Middle East. An important part of these supplies came from Caltex. In 1974 the group had restructured its operations in France and the previous arrangement whereby Caltex supplied crude in proportion to its ownership of the UIP Bordeaux refinery was annulled. UIP disappeared as a company and Caltex's interests in France were transferred to a 24 per cent interest in Elf-France (the French distribution and marketing subsidiary) together with an understanding to supply 4 million tons of Middle East oil per year at preferential prices on a long-term basis. Other supplies were obtained by means of the production service contract which had been pioneered in Iraq in exchange for a long-term supply agreement, and which had been subsequently extended to Iran on similar conditions.

By 1975, the year after the group had had such good results that for the first time in its history it paid a dividend to its principal shareholder, the state, the effects of the 1973 energy crisis were beginning to be felt. Although the immediate result of the changes in energy prices had been to put Elf–ERAP's receipts down and Aquitaine's up, gas prices were not fully aligned with oil, and sulfur prices were subject to government control by the Direction des Prix as a control against

Table 4. ERAP group oil production[28] (millions of tons)

	1970	1971	1972	1973	1974	1975	1976
ERAP	23.6	13.1	15.1	17.1	21.2	23.3	18.2
of which SNPA	2.5	1.8	1.7	1.9	2.7	2.7	–

*A premium crude has a low sulfur content and gives more products at the petrol end of the barrel.

inflation as sulfur was a basic chemical feedstock. Sulfur production, for much of the sixties an important contributor to SNPA's sales, had diminished in importance. Although SNPA production was essentially the entire part of French production, representing 6 per cent of world output, the market suffered from being dominated by US companies (with 40 per cent world output) and from occasional bouts of overproduction, particularly between 1969 and 1973, and by 1975 only 7 per cent of SNPA's sales came from this source.

Most importantly for the group as a whole, price controls were exercised on petroleum products and the group was also being squeezed on refining margins as demand and capacity utilization dropped. The effect of price controls was used by Guillaumat as further justification of the group's diversification policy away from hydrocarbons,[30] but the refinery problem was more intractable. In France, Elf-Aquitaine's principal, but not only market, good management seemed to have avoided the worst of the drop in refinery utilization which had affected many other European operators. Although in 1975 utilization was 69 per cent, in 1976 it was 80 per cent, and 1974 and the years before it had been the same or higher. Still, as refinery capacity exceeded sales in France, this gave the group considerable leeway to expand its market share under the Elf and ANTAR signs, where the group held 23.5 per cent of the market. Some of this share had been acquired by the takeover of independent operators, of which ANTAR and Caltex had been the most significant, but there was no question that a substantial part of the growth had been achieved by the direction of French oil policy in the market place by the government, and by good management by ERAP in taking advantage of the opportunities offered. There was further evidence of this in the expansion of sales outside France, i.e. steady in Italy, a price-controlled market like France, and improvement in Germany, a fiercely competitive market. In part the latter success was due to the purchase in 1975 of the Amoco European network which was strongest in Germany as well as the Occidental network which, while having some outlets in Germany, was particularly strong in the United Kingdom. Nonetheless the squeeze on refining was hurting the group, especially at a time when it had enormous capital requirements for developing the North Sea finds, Frigg and Ekofisk, and when its other markets for petrochemicals, sulfur, nickel and pharmaceuticals were not buoyant. In the face of their difficulties, Guillaumat, whose tenure of the presidency was extended in 1975 for two years

Table 5. Capacity, treatment, and sales volumes[31] (millions of tons)

	Refinery capacity		Volume treated		Finished products sold	
	France	Europe	France	Europe	France	Europe
1973	36.5	5.2	31.6	5.1	23.5	5.7
1974	35.3	6.6	31.6	5.4	20.6	5.5
1975	36.0	6.6	26.9	5.1	19.1	5.7
1976	36.0	6.6	29.1	5.0	20.9	7.0

Table 6. Finished products sold[32] (millions of tons)

	1974	1975	1976
France	20.55 (82%)	19.15 (77%)	20.81 (75%)
Germany	2.83	2.96	3.97
Benelux	1.15	1.12	1.22
Great Britain	–	0.90	1.04
Italy	0.68	0.75	0.75
	25.21	24.88	27.79

beyond his expected retirement, wanted to achieve the merger which had been his principal objective since 1971. It would bring considerable advantages to financing by sharing out SNPA's healthy cash flow over the group's activities as well as rationalizing into one company all the group's interrelated activities and thereby achieving managerial economies.[33]

The Merger

On January 9, 1976, the government and the managements of ERAP and SNPA all announced simultaneously that the final merger was to take place that year. The interests of the shareholders, public and private, and of the employees would be fully protected in the move, which was being made in the interests of efficiency and good management (according to the Elf-Aquitaine 1975 Annual Report). The assets of both groups were to be valued by two banks, the Chase Manhattan and the Société Générale, and then all the assets would be transferred to SNPA in exchange for shares. The shares would be in a newly named company, Société Nationale Elf-Aquitaine (SNEA), and would be held by ERAP acting purely as a holding company. The merger had to be approved by both the shareholders of SNPA and the Assemblée Générale. In the latter it met some opposition from the Left, who felt that the move was tantamount to denationalizing ERAP, but approval was given in July, with retroactive effect from January 1. With the state holding 70 per cent of the equity of the new company, no loss of control was likely, even though the state proposed to exercise its control with only 52 per cent. The remaining 18 per cent was to be frozen in a special account which might eventually be sold to the public.[34]

The merger marked the final step in the rationalization of the multiplicity of government-sponsored firms and interests in the petroleum sector which were created in the search for French oil. In thirty years BRP, starting as an interventionary policy body and holding company, had transformed itself into a managerially active investment and exploration company, then into a producing and refining group, BRP and UGP, into an operating company, ERAP, and finally into a fully integrated oil company, SNEA. With the consistent support of the French state and continuous involvement of one chief executive, Pierre Guillaumat, it was a considerable achievement.

In encouraging the merger the government may have been responding to criticism of its oil policy. In the face of the post-1973 economic difficulties and political pressures in the Common Market against overt assistance to the national companies and the consequential distortion to free competition, the government had been showing some reluctance to continue direct support. The original sales tax ('Fonds de Soutien') which had supported BRP came to an end in 1973 and the amounts received by ERAP from this source had been dropping since 1967. Complaints from the French independents that government policy was squeezing them out of the market had obliged the Commission to take notice and in addition the multinational groups operating in France were complaining that French price control and licensing policy was distorting the market.[35] This did not detract from the fact that the final rationalization of the state's oil interests into SNEA had much to commend it and consolidated under one management a strong group with considerable potential.

One of the successes of the group had been its diversification policy. The early accent on exploration had led the group to spread itself very wide and in 1977 with exploration and production in many corners of the globe the effort was justifying itself. As a direct by-product of this, SNPA had contributed what might be an important facet of activities in the future, an expertise in searching for minerals, and with an upturn in the world economies Le Nickel might prove to be less of a contentious venture. SNPA in fact was the start of much of the other diversified activities, which would be handled by the Elf–ERAP organization. The latter had been preoccupied first with the French market, and subsequently with developing an international network, in accord with French policy aims of having an important French presence in foreign oil markets equivalent at least to foreign presence in the French market. The external markets of the group were still only 23 per cent of sales, but the start was encouraging. The overall strength of the group was reflected in its financial performance, although so soon after what was undoubtedly a complicated merger, it was difficult to obtain a clear perspective. As an oil group, it compared favorably with other private oil operators (see Appendixes 9 and 10) and in size it was comparable with the American independents whose incursions into Europe in the fifties and sixties had now given way largely to operations of national companies. The operating figures for SNEA suggested that it ran its operations with efficiency comparable to the multinational companies and that it had as yet avoided being obliged by the government to undertake noncommercial activities in order to safeguard employment, although the investment in Le Nickel might be a bad augury, and the diversification around Pau was contentious for an oil company. Equally, the financial ratios showed no serious lack of profitability. A low self-financing ratio related in part to the large amounts devoted to investment and to the group's great indebtedness, the mark also of a national company; governments find loans easier to provide than equity. Finance had been a strong reason for the merger, as ever since 1973 the cash flow of Elf–ERAP had been poor and of SNPA very good. SNPA had always been more strongly internally financed, and consolidation enabled the benefits of this to be spread more widely (see Appendixes 11 and 12).

Even then Guillaumat was also of the opinion that if growth of Elf and SNPA was to continue, access to ever larger amounts of capital was needed. The growth was needed as neither SNPA nor Elf were yet self-sufficient in crude and they were working in a market which, in volume terms, would continue to grow even if the market share of the group did not change (see Appendix 6). At the same time the government was not following, at least in the eyes of Guillaumat, a clear energy policy.[36] De Gaulle had led a policy of oil under French control at any cost. Present policy, influenced by the 1973 energy crisis and the election in 1974 of President Giscard d'Estaing, a former Finance Minister, while accepting the dependence on oil, aimed to diminish the importance of oil in the energy balance. Thus the government's interest in divesting itself of its creation had coincided with Guillaumat's desire to diminish dependence on the government for the future growth of Elf. By taking in the private interests of SNPA, a way was opened to the private and international capital markets and, as one newspaper observed, 'the merger will give France a second integrated oil group of international dimension (the other being CFP) and it will be considered by the Government as a normal industrial enterprise responsible for its profitability and development'.[37]

The creation of SNEA, and a judgment on its future success, is difficult to separate from a judgment on the French oil policy, but in early 1977 SNEA gave every appearance of a solidly based group that could look to the future with some confidence. 1977 was the year in which Pierre Guillaumat retired, and to him much credit could be given for leading or guiding over a period of thirty years such a successful metamorphosis from administrative body to international oil company. In a country where administrative form is strong and inflexible, drive was needed to overcome the many administrative problems posed by such a change.

The Post-Guillaumat Era

In August 1977, Albin Chalandon took over as SNEA's new chief executive officer. Chalandon, a former Minister of Equipment, was a political appointee with limited background in the energy field. This was in line with governmental thinking which wished to see the government's role diminished in SNEA and give the company more independence from its principal owner. In France, paradoxically, such an orientation was thought to be better accomplished by an executive who was primarily a politician and, given the French political and social environment, this is indeed likely to be the case.

Guillaumat was a tough act to follow. The task was not made easier by yet another oil 'crisis' in 1979, partly due to the Iranian oil cutbacks, when crude prices rose by some 60–65 per cent. Yet, because of SNEA's strong crude position, the company benefited from the increases in crude oil prices. Chalandon, to gain the confidence of long-time SNEA associates, created an Office of the Chief Executive comprising himself, the two executive vice-presidents, and a newly appointed group executive director. He, himself, concentrated his efforts to do battle with the Minister of Industry, André Giraud, to increase the controlled product prices, to allow SNEA to rationalize its refinery and marketing operations, or, if this was not allowed, to apply for governmental subsidies. Chalandon's emphasis, more than that of his predecessor, was to make SNEA more profitable than it had been. In this, partly because of internal measures and partly because of SNEA's strong crude position, he has been successful.

Internal measures included the virtual integration of the Elf and ANTAR activities thus achieving economies of scale. In the process, unprofitable refining operations and distribution outlets were closed. Elf also reversed its policy of giving traders generous discounts and, as a matter of deliberate policy, reduced its market share (from 23.3 per cent in 1977 to 22.7 per cent in 1978). 1979 also witnessed another quasi-internal measure, namely, Standard Oil of California and Texaco sold their combined 19 per cent in Elf-France to SNEA. (See Appendix 13.) The two companies, though, maintained their long-term crude-supply agreements, thus not affecting the important crude availability to SNEA.

SNEA continued its ambitious exploration program; the company is, as of 1980, among the worldwide leaders in searching for oil and gas and, in this respect, is ahead of companies with sales far in excess of it. While SNEA is exploring on every continent it has an announced preference to politically stable areas even though these tend to be more expensive than, say, the Persian Gulf. Because of SNEA's past oil discoveries – and the increase in crude oil prices – the 1978 and 1979 financial returns showed a marked improvement over 1977. Thus SNEA continued on its path of increased financial independence from the government which was both the desire of the government and Guillaumat in creating SNEA in the merger of 1976.

Appendix 1. Production of Crude Oil in the Franc Zone[a] [38] (millions of tons)

Region	Recapitulation (Total not kept of breakdown)				France								
					Alsace		Paris Basin				Aquitaine		
Position	France	Algeria	West Africa	Total	Pechelbronn	Scheibenhard Eschau, etc.	Chailly St. Firmin Chuelles Trigueres Chateau-Renard	St. Martin Gisy etc.	Coulommes	Villemer Brie etc.	Parentis Cazaux Mothes Lugos, etc.	Lacq Superieur	Total
Operation					Pechelbronn	Prifa	CTP	Cupusip	Petrorip	IHAP	Essorip	SNPA	
Before 1946	2,657.8	33.0	–	2,690.8	2,671.0		23.4	–	–	2.4	–	–	2,696.8
1946	51.6	0.2	–	51.8	51.4	–	0.1	–	–	0.1	–	–	51.6
1947	49.9	0.1	–	50.0	49.3	–	0.1	–	–	0.5	–	–	49.9
1948	51.7	0.1	–	51.8	50.0	–	0.1	–	–	8.6	–	–	58.7
1949	57.9	0.3	–	58.2	57.0	–	0.4	–	–	0.4	–	0.1	57.9
1950	127.8	3.7	–	131.5	61.6	–	0.8	–	–	0.1	–	65.3	127.8
1951	290.8	7.3	–	298.1	56.8	–	0.1	–	–	–	–	233.9	290.8
1952	350.3	46.0	–	396.3	51.4	0.7	1.0	–	–	–	–	297.2	350.3
1953	367.4	84.9	–	452.3	45.8	13.0	1.3	–	–	–	–	307.3	367.4
1954	505.2	75.8	–	581.0	40.1	25.4	0.5	–	–	0.2	131.5	307.5	505.2
1955	878.4	57.8	–	936.2	35.7	14.6	0.3	–	–	0.2	576.5	251.1	878.4
1956	1,263.6	33.5	–	1,297.1	32.8	14.4	0.4	–	–	–	1,033.8	177.8	1,259.2
1957	1,410.5	21.3	173.2	1,605.0	27.7	32.9	0.4	–	–	–	1,223.2	126.3	1,410.5
1958	1,386.3	428.9	504.8	2,320.0	26.9	41.6	2.0	–	11.5	–	1,207.9	101.4	1,391.3
1959	1,610.2	1,232.4	753.3	3,595.9	22.8	45.8	38.0	17.4	243.2	12.7	1,232.0	100.1	1,690.0
1960	1,976.5	8,599.7	853.6	11,429.8	22.3	41.5	110.1	133.7	178.3	56.8	1,345.4	88.4	1,976.5
1961	2,163.4	15,664.4	879.6	18,707.4	20.6	33.1	181.3	141.9	122.1	70.1	1,503.2	91.1	2,083.4
1962	2,370.2	20,497.7	950.9	23,818.8	14.5	29.5	262.0	110.7	110.8	41.4	1,676.4	85.9	2,331.2
1963	2,522.2	23,646.4	998.9	27,167.5	6.2	27.8	278.1	85.2	92.7	75.8	1,870.0	86.4	2,532.2
1964	2,845.5	26,231.2	1,141.9	30,218.6	5.4	25.0	320.1	69.4	93.6	73.8	2,176.8	81.9	2,895.5
1965	2,987.8	26,025.0	1,335.7	30,348.5	–	24.2	122.1	60.9	79.5	48.4	2,155.6	86.7	2,987.8
1966	2,932.0	33,257.2	1,509.5	37.698.7	–	21.1	301.4	44.9	74.0	111.7	2,336.9	101.9	2,912.0

Position	North Algeria[b]		Algeria (Sahara)										West Africa[c]	
	Oued Gueterini	Guerel ONK	Hassi-Messaoud (N) Haoud-Beanaoui	Hassi-Messaoud (S) Nezla**	El Gassi El Agrer	Ohanet (N) Askarene Guelta, etc.	Zaizaitine	Eguele-Tiglentourine El Ader Larache Ohanet (S) etc.	Tan Emellel (S)	Gassi-Touil	Rhourde-El-Paguel	Rhouroe Nouss	Olvers Gabon	Pt. Indienne Congo
Operation	Cared	SN-Repal	CFP (A)	SN-Repal	SNPA	CTP	Creps	Creps	Eura-tree	Copeta	Sinclair	El Pago	Spaff	Spaff
1957	13.4	–	–	5.6	–	–	–	4.1	–	–	–	–	171.9	–
1958	3.2	–	210.2	208.0	–	–	–	7.5	–	–	–	–	504.4	–
1959	3.1	–	553.0	655.1	–	–	–	20.6	–	–	–	–	153.3	–
1960	4.2	0.4	2,373.1	4,271.4	73.5	–	1,447.8	484.3	–	–	–	–	800.1	61.8
1961	3.6	4.4	3,321.6	4,877.7	466.7	288.2	5,229.4	1,472.8	–	–	–	–	714.5	102.9
1962	5.2	5.0	3,670.5	5,506.8	750.1	587.3	7,307.1	2,634.1	31.6	–	–	–	821.0	123.4
1963	4.1	5.1	4,382.4	6,562.8	1,116.9	856.2	6,686.9	3,799.1	38.3	11.9	182.1	–	889.7	109.2
1964	4.1	1.2	4,938.1	7,318.6	714.4	924.2	6,591.7	4,395.5	102.5	230.2	1,010.7	–	1,058.8	83.6
1965	3.1	–	5,841.5	7,319.1	581.4	1,198.2	5,317.9	4,535.3	110.8	637.2	970.9	–	1,264.4	71.3
1966	3.5	3.4	6,246.4	8.489.4	1,556.9	1,244.2	3,624.5	5,185.9	94.6	3,473.5	3,202.3	182.6	1,447.0	62.5

[a] ERAP shareholding.
[b] Up to 1956, see Recapitulation Table above.
[c] There is room to add that production from the Diam Hiadé field of the SAP at Senegal is 1.7 in 1960, 2.2 in 1961, and 0.5 in 1962 (compiled in the Recapitulation of West Africa above).
*Production compiled as 5.8 in 1965 and 19.8 in 1966.
**Production compiled as 17.3 in 1966.

Source: *Direction des Carburants.*

Appendix 2. Financing of Elf-Aquitaine Consolidated[39]

BRP + RAP + SNPA from 1961 to 1965 – ERAP + SNPA from 1966 to 1975 – ERAP + SNEA in 1976 – SNEA not including ERAP in 1977 and 1978 (millions of francs)

	1961	1962	1963	1964	1965	1966	1967	1968	1969
Requirements	948	749	1,367	1,203	1,382	1,705	1,821	1,789	2,130
Private capital	—	25	41	46	—	29	34	38	55
Dividends to shareholders	–20	–55	–188	–193	–123	–89	–102	–166·	–179
State capital	213	208	202	181	262	358	382	364	305
Dividends to state	—	—	—	—	—	—	—	—	—
Net borrowing	34	–64	–135	–156	23	166	307	128	203
S/TOTAL									
External financing	227	113	–80	–122	162	464	621	364	384
Internal financing	851	809	1,309	1,367	1,127	1,280	1,420	1,683	2,273
Changes in working capital	–130	–173	+138	–42	+93	–39	–220	–258	–527

	1970	1971	1972	1973	1974	1975	1976	1977	1978
Requirements	2,900	2,498	2,850	3,098	5,854	9,088	8,710	8,523	7,065
Private capital	186	58	58	97	46	28	171	102	57
Dividends to shareholders	–177	–57	–122	–140	–177	–235	–272	–299	–385
State capital	267	214	230	200	100	100	—	71	50
Dividends to state	—	—	—	—	—	—	–50	–65	–65
Net borrowing	119	241	223	785	1,081	2,943	3,469	2,589	–450
S/TOTAL									
External financing	395	456	389	942	1,050	2,736	3,368	2,398	–793
Internal financing	2,510	2,361	2,011	2,634	4,303	5,496	5,187	5,775	7,700
Changes in working capital	–5	–319	+450	–478	+501	+856	+155	+350	+158

1977–1978: Provisional forecast

Appendix 3. Financial History of SNPA During its 34 Years[40] (millions of francs)

	1942–49 (8 yrs)	1950–54 (5 yrs)	1955–60 (6 yrs)	1961–64 (4 yrs)	1965–69 (4 yrs)	1970–73 (4 yrs)	1974–75 (2 yrs)	Total (34 yrs)	Total (per cent)
Total requirements	28	136	1,418	1,493	2,946	2,994	5,491	14,506	100.0
Financing									
(A) Public organisms									
Capital	11	4	176					191	1.3
Subsidies		72	28					100	0.7
Long-term loans from BRP	9	3	233					245	1.7
S/TOTAL A	20	79	437					536	3.7
(B) Private financing									
Capital	8	2	158		2	112	5	287	2.0
Loans			467	25	394	518	1,618	3,022	20.8
S/TOTAL B	8	2	625	25	396	630	1,623	3,309	22.8
(C) Self-financing		55	356	1,468	2,550	2,364	3,868	10,661	73.5
Gross dividends (by period)		2	1	209	453	519	334	1,518	10.5

Appendix 4

Natural gas production and sales[41] (millions of cubic meters)

	Production		Sales
	ERAP	SNPA	ERAP + SNPA
1946	110	–	100
1950	246	–	227
1955	274	–	256
1960	300	4,116	2,883
1965	216	7,678	5,048
1970	187	10,063	6,880
1974	97	11,015	7,625

Note: The difference between production and sales volumes is due to products (principally sulfur) extracted by treatment.

Natural gas consumption[42] (billions of therms)

	Home production[a]	Algerian imports	Dutch imports
1960	26.3	–	–
1965	45.8	1.9	–
1970	64.7	5.5	26.0
1974	69.2	21.2	76.0

[a]99.96 per cent produced by ERAP.

Appendix 5. Refineries, ERAP and ANTAR Ownership, 1975[43]

	Share (per cent)	Capacity (millions of tons)	Principal other partners
ERAP			
Ambès, Bordeaux	100	2.1	–
Feyzin, Lyon	100	8.8	–
Grandpuits, Paris	100	4.3	–
Gargenville, Paris	100	6.0	–
Hanconcourt, Metz	9	4.0	CFR, Esso
Reichstett, Strasbourg	10	4.0	Shell, Mobil
Klarenthal, Germany	10	2.4	CFP, ANTAR, Saarbergwerke
Spire, Germany	100	3.2	–
Madagascar	27	0.75	Esso, Caltex, Shell, BP, SPM, CFP
Martinique	25	0.55	Shell, Esso, CFP, Texaco
Cambodia	35	0.60	Cambodia institutions
Senegal	30	0.65	Shell, BP, CFP, Esso, Mobil, Texaco
Ivory Coast	25	2.00	Mobil, Shell, CFP, BP, Texaco
Gabon	18.75	0.95	CFP, Mobil, Shell, Texaco, BP
ANTAR			
Donges, Nantes	100	8.3	–
Vern-sur-Seiche, Rennes	100	1.5	–
Valenciennes, Lille	100	3.6	–
Herrlisheim, Alsace	14	6.0	CFR, BP
Klarenthal, Germany	10	2.4	CFP, Elf, Saarbergwerke

Appendix 6

Elf market share[44] (per cent)

	1966	1968	1970	1972	1974	1975
Petrol	8.93	10.95	12.04	13.07	12.97	12.96
Gas oil	11.49	13.17	12.11	12.47	13.01	13.04
Domestic fuel	13.29	13.26	14.50	13.32	12.37	12.40
Light oil	13.03	12.59	13.33	16.33	19.03	21.26
Heavy oil	11.77	14.04	14.87	16.65	14.95	16.05
All products	11.80	12.97	13.99	14.46	13.63	13.99
Millions of tons	4.9	6.9	9.8	12.4	12.4	11.5
Lubricants	NA	NA	NA	7.91	9.31	10.25
Bitumens	NA	NA	NA	10.06	10.60	10.39

ANTAR market share (per cent)

	1970	1972	1974	1975
Petrol	9.54	9.44	10.66	10.46
Gas oil	11.81	11.27	11.54	12.01
Domestic fuel	9.29	8.48	8.65	8.87
Light oil	8.53	8.68	9.61	9.80
Heavy oil	7.95	6.73	8.31	8.24
All products	9.07	8.23	9.09	9.23
Millions of tons	6.3	7.1	8.1	7.6
Lubricants	NA	10.52	10.51	11.19
Bitumens	NA	5.93	6.70	6.93

Appendix 7. SNPA[45]

Exploration investment – geographic distribution (per cent)

	1962	1970	1975
France	79.3	9.4	19.0
Europe (outside France)	8.6	4.4	17.0
Africa and Middle East	10.4	30.6	30.0
North America	0.0	34.0	21.0
Australia and Far East	1.7	21.6	13.0
Gas production (billions of cubic meters)			
France	4.42	6.81	7.28
Overseas	–	0.14	1.88
Oil production (millions of tons)			
North America	–	0.99	0.98
Africa	0.38	1.47	1.44
France	0.09	0.06	0.07
Norway	–	–	0.25
Sales distribution (per cent)			
Gas	52	28	27
Sulfur	23	18	7
Oil	22	22	32
Petrochemicals	3	32	13
Pharmaceuticals	–	–	19
Coal	–	–	2

Appendix 8. Elf Crude Oil Production[46] (thousands of tons)

	1970		1975	
	Elf	SNPA	Elf	SNPA
Italy	–	–	31	–
Norway	–	–	454	227
France	235	58	220	66
Algeria	16,489	1,041	5,238	–
Tunisia	28	167	891	809
Libya	136	272	83	164
Gabon	4,134	–	9,739	–
Congo	19	–	1,179	–
Nigeria	–	–	1,704	–
Canada	–	1,046	–	1,583
United States	–	–	–	109
	21,041	2,584	19,539	2,958

Appendix 9. SNEA, Its Private Peers and Superiors[47]

Comparative operating data, 1975

	Atlantic Richfield	Continental	SNEA	Phillips	Texaco	Socal
Oil production (millions of tons)	30	27	23	20	188	155
Refining production (millions of tons)	35	23	30	28	139	105
Gas production (billions of cubic meters)	19	16	10	17	42	15
Personnel (thousands)	28.1	44	23.4	30.5	75.2	38.8
Investment (million $)	1,751	747	1,808	694	1,546	1,253
Assets (million $)	7,365	5,185	8,568	4,545	17,262	12,898
Turnover (million $)	7,308	7,253	6,841	5,134	24,508	16,822
Total profit (million $)	350	365	334	342	847	773
Cash flow (million $)	925	807	1,047	719	1,753	1,551

Appendix 10. Comparative Operating and Financial Ratios, 1975[48]

	Atlantic Richfield	Continental	SNEA	Phillips	Texaco	Socal
Debt ratio[a] (per cent)	27	25	38	24	18	16
Self-financing ratio[b] (per cent)	41	84	50	85	80	92
ROI[c] (per cent)	10	17	14	18	10	13
Sales/employee (million $ per thousand)	260	165	292	168	326	434
Assets/employee (million $ per thousand)	262	118	366	149	230	332
Investment/employee (million $ per thousand)	62.3	17	77	23	21	40
Profits/sales (per cent)	4.7	4.6	4.8	6.7	3.4	4.6
Profits/assets (per cent)	4.7	6.4	3.9	7.5	4.8	6.0
Cash flow/sales (per cent)	13.0	11.1	15.3	14.0	7.2	9.2
Oil production/employee (million tons per thousand)	1.06	0.61	0.98	0.66	2.50	3.89
Refined products/employee (million tons per thousand)	1.25	0.52	1.28	0.92	1.85	2.71
Gas production/employee (billion cubic meters per thousand)	0.68	0.36	0.43	0.56	0.56	0.39

[a]Debt ratio = long-term debt/owners equity.
[b]Self-financing ratio = cash flow − dividends/investments.
[c]ROI = net profit/net assets.

Appendix 11. Group Elf-Aquitaine Consolidated Financing as a Percentage of Requirements[49]

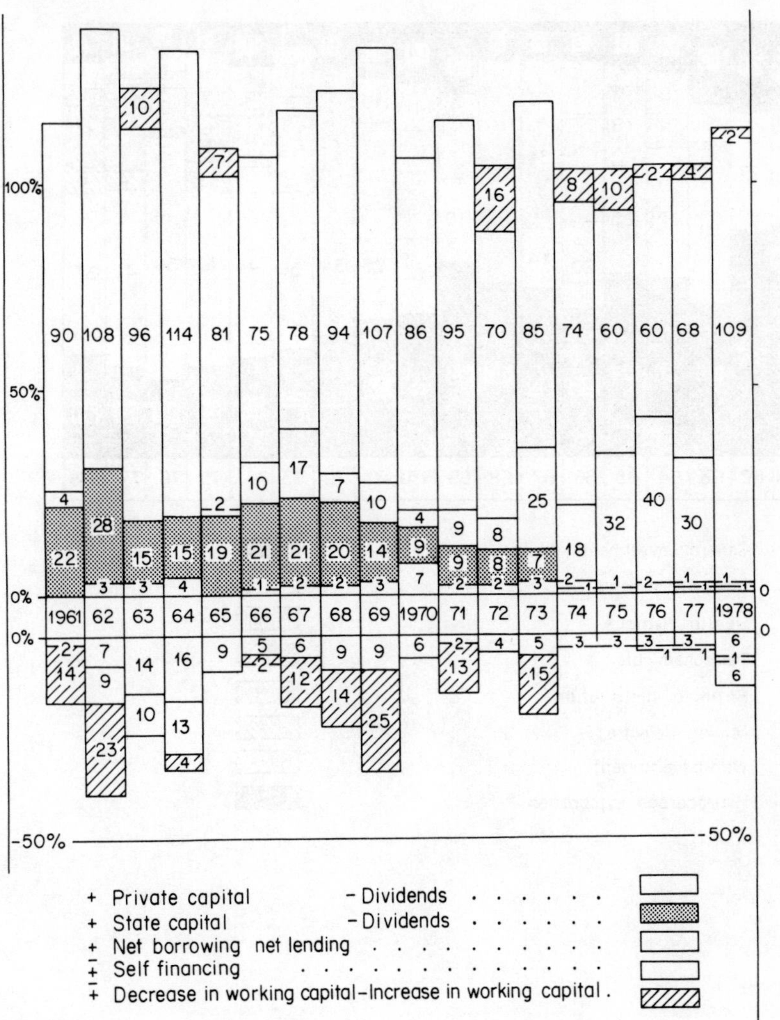

Appendix 12. Group Elf-Aquitaine Consolidated Distribution of Investments[50] (per cent)

General investments, financial and others

Other investments DGC

Health, hygiene

Petrochemicals

Refining, distribution

Mining activities

Well development

Hydrocarbon exploration

Appendix 13

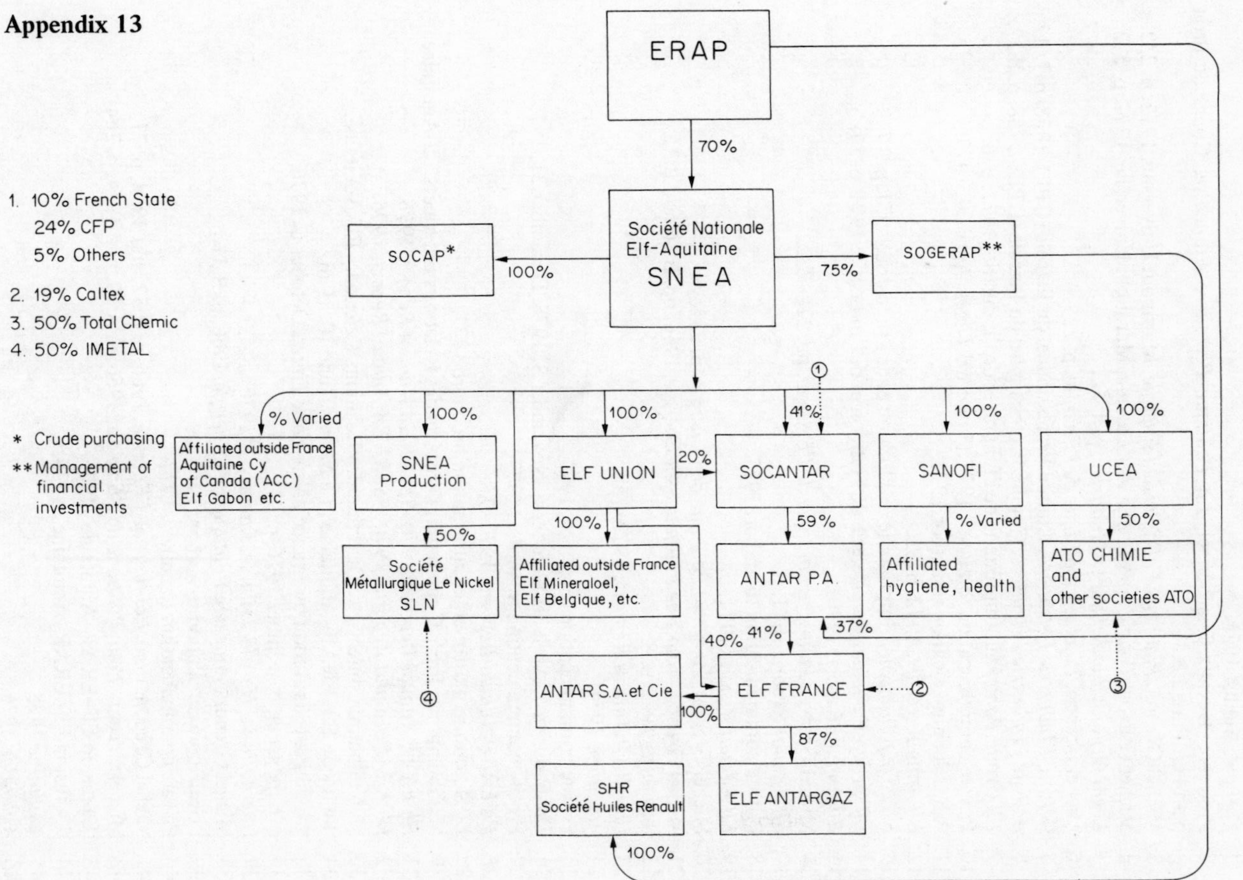

1. 10% French State
   24% CFP
   5% Others
2. 19% Caltex
3. 50% Total Chemic
4. 50% IMETAL

* Crude purchasing

** Management of financial investments

References

1. *Fortune*, August 1976.
2. Professor Devaux-Charbonnel, *La Politique Pétrolière Française*, Cours de droit pétrolier, Paris I et II.
3. Michel Grenon, *Pour une Politique de l'Énergie*, Marabout Université, 1972, p. 222.
4. Michel Grenon, *Pour une Politique de l'Énergie*, Marabout Université, 1972, p. 222.
5. *Legal Organization of the Oil Industry*, p. 4.
6. Michel Grenon, *Pour une Politique de l'Énergie*, p. 225.
7. *Activité l'Industrie Pétrolière*, Comité Professionnel du Pétrole (CPP), 1968, p. D 17.
8. *Activité l'Industrie Pétrolière*, Comité Professionnel du Pétrole (CPP), 1968, p. B 32.
9. *Elf Group Pamphlet*, Imprimerie de l'Edition et de l'Industrie, 1971, p. 6.
10. Private communication to author 1977; *Financial Times*, April 29, 1977.
11. Private communication to author 1977.
12. *Elf Annual Report*, 1972, p. 32.
13. *Elf Group Pamphlet*, Imprimerie de l'Edition et de l'Industrie, 1971, p. 17.
14. *Elf Group Pamphlet*, Imprimerie de l'Edition et de l'Industrie, 1971, p. 17.
15. *Elf Annual Report*, 1970.
16. Michel Grenon, *Pour une Politique de l'Énergie*, pp. 228–232.
17. *Elf Annual Report*, 1971.
18. Private communication to author, 1977.
19. *Elf Annual Report*, 1971.
20. Michel Grenon, *Pour une Politique de l'Énergie*, p. 232.
21. Michel Grenon, *Pour une Politique de l'Énergie*, 1966, pp. B 16 and B 31.
22. *Aquitaine 1972*, p. 8.
23. *Elf ERAP Annual Reports*, 1970–75.
24. *Aquitaine 1974*.
25. Private communication to author, 1977.
26. *The Economist*, April 28, October 6, November 3, 1974, December 21, 1975.
27. Private communication to author, 1977.
28. *Elf ERAP Annual Reports*, 1970–75.
29. *Elf Aquitaine*. Post-merger information brochure 1977.
30. M. Guillaumat, Press Conference, Direction des Relations Extérieurs Elf-Aquitaine.
31. *Elf ERAP Annual Reports*, 1970–75; *SNEA Annual Report*, 1976.
32. *Elf ERAP Annual Reports*, 1970–75; *SNEA Annual Report*, 1976.
33. M. Guillaumat, Interview, Direction des Relations Extérieurs Elf-Aquitaine.
34. 'La fusion Elf–ERAP–Aquitaine', *Le Monde*, June 10, 1976.
35. 'Mobil Protests Action by France', *New York Times*, October 6, 1976.
36. *L'Expansion*, November 1976, p. 97.
37. *International Herald Tribune*, January 9, 1976.
38. Michel Grenon, *Pour une Politique de l'Énergie*, 1966, p. B 31.
39. Private communication to author, 1977.
40. Private communication to author, 1977.
41. Michel Grenon, *Pour Politique de l'Énergie*, 1966, p. B 32, and 1974, p. 77.
42. Michel Grenon, *Pour Politique de l'Énergie*, 1966, p. A 17, and 1974, p. 264.
43. 'La fusion Elf–ERAP–Aquitaine', 1975.
44. 'La fusion Elf–ERAP–Aquitaine', 1975.
45. *Aquitaine 1976*.
46. *Aquitaine 1976*.
47. Computation by author based on data in private communication, October 1976.
48. Computation by author based on data in private communication, October 1976.
49. Private communication to author 1977.
50. Private communication to author 1977.

CHAPTER 5

ENI: Italian National Oil Company

Ente Nationale Idrocarburi (ENI) is the Italian State Energy Corporation with interests particularly in oil and gas exploration, production, transport, refining and marketing, nuclear energy, chemicals, mechanical engineering, pipeline construction, together with an assortment of other activities. In the field of hydrocarbons alone, which today account for 82 per cent of sales, its activities are as varied and widespread as those of a multinational oil company, and its size places it eleventh among the oil companies of the world and makes it the largest national oil company after the National Iranian Oil Company (NIOC) and Compagnie Française des Pétroles (CFP). (NIOC is not an integrated company; it is primarily a seller of crude oil.) ENI's work is carried out by nine subsidiary companies of which AGIP, S.p.A., responsible for oil and natural gas exploration and production, and the marketing of petroleum products, is the best known, and was the founder company of the group. Perhaps better than any other national oil company ENI illustrates the problems that face a group which has to operate in a commercial field but uses criteria other than purely commercial ones for its management. The boundary between politics and commerce, if it exists, runs straight through ENI.

The Opening Years

Unlike the other Western European countries, Italy never had a plentiful source of cheap energy to encourage industrial development in the nineteenth century. Coal prices were twice those of France, Britain, and Germany. Despite being an early user of oil products – the streets of Genoa and Parma in 1802[1] were lit with oil from wells in the River Taro basin – the Italian government could not interest private operators in Italian oil exploration and production and therefore took matters into its own hands.

Azienda Nationale Generale Italiani (AGIP) was founded in 1926 to conduct Italian exploration, refining, and marketing activities at home and abroad. Precedents for national oil companies already existed in Britain with British Petroleum (BP) – one of the multinationals – and in France with CFP. A common motivation for all three companies was national control of oil supplies in a domestic market already dominated by the Americans, but AGIP, unlike CFP, from the start had no oil supplies of its own.

AGIP soon acquired a few oil companies in Rumania, made several long-term agreements for Albanian oil and in 1930 took a minority interest in British Oil Developments Ltd. (BOD), an Iraqi oil company. By 1935, AGIP had gained complete control of this company, but exploration successes were limited and Italian–British tension and Mussolini's foreign exchange needs obliged AGIP to sell the company in 1936.[2] Undramatic and relatively unsuccessful as these activities were, they helped establish a cadre of experience which was to be of use to AGIP in its activities after World War II.

Government action did not stop with the founding of AGIP. In 1932 special status had been given to Italian refineries and in 1936 Azienda Nationale Idrogenazione Combustibile (ANIC), with a 25 per cent AGIP interest, was formed to take advantage of this. By this time, AGIP had created an efficient distribution network for petroleum products, meeting 25 per cent of Italian consumption, but including some preferential governmental outlets, not now held. However, then as now, AGIP's supply of crude oil was insecure, and less than its requirements.

The cessation of hostilities in 1945 brought to the management of AGIP, and in particular to its exploration activities in the Po Valley, a Christian Democratic resistance fighter, with whom the name of ENI will always be associated, Enrico Mattei. Despite official disillusion with the results and costs to date, Mattei continued with the search for energy supplies in the Po Valley, perhaps on the basis of a promising geological study made before the war using a new technique.[3] An activist by nature, Mattei quickly made known his concept of the role for AGIP (of which he was vice-president) to free the Italian market from foreign domination. He saw world oil markets controlled by what he scornfully called the 'Seven Sisters' (Esso, Texaco, Mobil, Gulf, Standard of California – all of the United States; British Petroleum and Royal Dutch/Shell – of Britain and the Netherlands), and Mattei's objective was to break into this oligopoly and establish AGIP as a world force. For this, he had the tacit backing of the Italian political world, in which, as an important local (North Italy) Christian Democrat, he had both influence and contacts. However, AGIP's continuing weakness was its lack of any really substantial supplies of oil of its own.

In 1949, oil and gas were struck in Cortemaggiore in the Po Valley. Although the gas finds turned out subsequently to be substantial and the oil did not, from the start Mattei capitalized on the find to promote what he saw as AGIP's role, even to the extent of implying at AGIP's petrol stations that all the petrol was of Italian origin, with the legend 'Supercortemaggiore La Potente Benzina Italiana'. Mattei stampeded Parliament into granting AGIP exclusive control of the underground resources in the area, rushed to build a monopoly gas distribution network, and priced the gas in relation to fuel oil rather than production cost. By these two last moves, he established the basic strength of AGIP for years to come. By 1950 Mattei was in complete control of AGIP. His corporate power, his political influence, and the public support generated by the Po gas strike enabled him to achieve his most important coup in 1953 with the creation of ENI, of which he was made president.[4] (See Appendix 1 for initial organization and capital.)

ENI – The Mattei Era 1953–62

Established on February 10, 1953, ENI was to be the holding company for the various state companies operating in the petroleum sector. Its charter of incorporation obliged it to follow ventures of national interest in the fields of basic petroleum, natural gases, chemical and nuclear fields, together with related mining activities. The right to exclusive exploration and transportation of gas and petroleum in the entire Po Valley was to become the cornerstone of its success, but the company was otherwise expected to compete on an equal footing with the private companies operating in the oil sector.

The gas monopoly granted ENI and the lack of strong political guidance from the Italian government gave Mattei almost complete freedom to decide who should be supplied with gas and for how much. ENI thus was able to build up an extremely profitable business whose large cash flows further reinforced the independence of the group by making its investment requirements to a large extent self-financed. Mattei's influence was further extended by his use of ENI's funds to buy an advertising agency, a news service, and a Milan newspaper, *Il Giorno*, which put his own point of view against those newspapers controlled by other industrial groups hostile to the growing power of ENI. Although obliged to give up his seat in the Chamber of Deputies by the passing of a law in 1953 preventing MPs' holding positions in government corporations (intended not least to limit Mattei's power), such was his political strength that soon his freedom of action on the domestic scene was translated to the area of foreign affairs where for years he strongly influenced Italian foreign policy in energy matters.

Mattei continued to exercise political power from the corporate boardroom and as ENI grew his contribution to Italian foreign policy became more influential. However, one fact remained constant: ENI, or more particularly, AGIP and ANIC, were almost totally dependent on non-Italian controlled supplies of crude oil. AGIP had a longstanding arrangement of supply with British Petroleum (formerly as Anglo-Persian) and ANIC with Esso, both the result of post-war reorganization. Mattei's first contact with foreign oil policy followed the Iranian nationalization of BP. When the Shah asked Western companies to form the Iranian Oil Company to take over BP's operations, AGIP's request for a 1 per cent participation was spurned by the seven major companies even though AGIP had been loyal to BP throughout, and had had previous interests in the area with BOD. But worse was the fact that five purely domestic American companies were given shares in the consortium. This incident only served to reinforce Mattei's conviction that AGIP must find independent sources and he devoted all his energies to this, bringing to bear his own ideas of the benefits required for the Italian people and for the producing states, interests which he felt were ignored by the major oil companies.

The decade from 1953 to 1962 was one of hectic expansion for ENI. The marketing of AGIP products had been much improved before its absorbtion into ENI, and by 1955 AGIP had become renowned for its superbly equipped and serviced petrol stations. In the next year ENI started a chain of associated motels which enhanced Mattei's public image and provided useful support against frequent political attacks. The attackers accused Mattei of wielding too much unbridled power, and of building petrol stations which had more in common with monumental cathedrals than service outlets. To limit his power, Mattei's enemies managed to have a law passed in 1957 preventing further joint ventures with private enterprise. The law was repealed after Mattei's death.

As a result of reorganizing a bankrupt engineering firm, NUOVO PIGNONE, which was forced on ENI by the Italian government in 1956, ENI was competing in all the producing fields with its own equipment. Later, in 1961, ENI was similarly obliged to take over a textile, Lanerossi, which, unlike NUOVO PIGNONE, was to prove a financial albatross for many years, right up to present times. Also in 1956, ENI built a petrochemical complex at Ravenna with the technical aid of Phillips Petroleum, created AGIP NUCLEARE, formed a contract engineering subsidiary SNAM PROGETTI as an offshoot of the pipeline subsidiary company Societa Nationale Metanodotti (SNAM), and formed a financial petrochemical investment group, SOFID.

In 1957, Mattei negotiated a joint exploration and production agreement with Iran and Egypt. Previous agreements between the oil companies and the producing countries had been on a royalty basis of 50–50. Mattei's arrangement was about 75–25 in favor of the producer country and at the same time also offered the producer country a participation stake in its own oil. Other oil companies soon had to follow ENI's precedent of producer country participation.[5]

ENI entered into similar agreements with other governments, mostly in Africa, in exchange for various obligations such as technology transfer, joint investment, or for barter arrangements. The preferred policy of dealing on a government to government basis derived from a longstanding philosophy of Mattei's aided by his powerful reputation that made his name almost synonymous with the Italian government. Generally, ENI's market initiatives were, in purely marketing terms, somewhat too early, but this was due to the perceived need to be ahead of the majors in order to be free of them. By the end of 1959 ENI had significant investments in Austria, Germany, Switzerland, Sudan, Morocco, and Tunisia, had drilled the first European offshore well, and had initiated a price war for petrol in Italy.

In the same spirit of independence, and forever looking for new opportunities and sources of crude, ENI entered into relations with the Algerian revolutionary government in an effort (partially successful subsequently) to break into the French Sahara, a move which did little to improve Italian–French relations. In 1961, ENI signed a long-term crude oil contract with the USSR, an agreement that typically included the exchange of industrial commodities and services for an oil supply. The agreement has continued and been extended to the supply of gas. This was the first major supply agreement made totally out of the control of the majors, and it considerably affected American views of East–West relationships and sensibilities.

The particular style of Mattei, autocratic, ambitious but with considerable flair, undoubtedly had much to do with the rapid expansion of ENI, not least because of the speed of decision making, but by the sixties the strains were beginning to tell. In 1957 Mattei had asked an American management consulting group to help him reorganize the group,[6] but that did not affect his pervasive influence. Some of his investment initiatives, particularly the refinery in Germany, were beginning to look distinctly premature, and he was also realizing that his one-man war against the Seven Sisters could not continue indefinitely. Already by the time of the Russian deal he was making overtures to Esso, having essentially decided that ENI had made its mark. By the beginning of the sixties, ENI was by any measure an integrated oil company of world scale, and despite some problems, was unquestionably a force to be heeded.

On October 27, 1962, Enrico Mattei died when his private plane crashed in a storm. Under his guidance, ENI, between 1954 and 1963, trebled its turnover, doubled gas production, quadrupled employment, changed its oil production from effectively zero to four million tons per annum, and doubled its pipeline network.

The Consolidation Period 1962–72 Under Cefis

On the death of Mattei, the then vice-president, Boldrini, took office, but there was in fact only one man who was capable of stepping into Mattei's position, Eugenio Cefis, a war-time friend and adviser, and the only person in whom Mattei confided to any significant extent. He had left ENI in 1959 after a quarrel with the Ministry of State Holdings. Cefis, taking most of the executive power in his hands, became vice-president, and president in 1967.[7] A powerful personality and a formidable politician like Mattei, Cefis shunned publicity, leaving public duties to Boldrini. He saw his objective at ENI to be fourfold: to sort out the corporation's difficult financial situation, to streamline the organizational sprawl, to clarify ENI's purpose and strategy, and to build up an effective high command. In addition, Cefis also continued the *rapprochement* that Mattei had already initiated with the majors. However, Cefis did not modify the aim of ENI that his predecessor had established. He continued Mattei's vision of making ENI the world's largest state oil concern and the dominant enterprise on the Italian market.

In 1962, ENI announced refinery ventures in the Congo, quickly followed by a similar agreement with Tanzania in 1963 and a large investment in Ghana. But 1963 also saw the official result of drawing closer to Esso, which signed a five-year, 10-million-ton supply agreement with ENI at posted prices[8] (not 20 per cent below as some observers have suggested[9]). As with ENI's other crude-supply agreements, 50 per cent of the f.o.b. price was paid with machinery and services from ENI affiliates. In the same year, the Soviet supply agreement was renewed for a further six years, giving Italy a larger quantity of supply, although in proportion to total imports, the amount was smaller.

The search for crude continued, but ENI's efforts at exploration had been notably unsuccessful and in 1964, ENI came to a wide-ranging agreement with Gulf, involving the purchase of the Ragusa (Sicily) crude field and the supply of 5 million tons of crude over a six-year period at 27 per cent below the posted price.[10] To effect this purchase, ENI was obliged to raise a loan from the Mellon interests, who were, and are, substantial Gulf shareholders.

By 1965, the crude-supply situation had begun to brighten as the Tunisian field had been declared commercial, and finds were being reported in Nigeria and Libya. Esso continued as an obliging partner and signed a twenty-year natural gas supply contract of 3 billion cubic meters per year for Libyan LNG.[11] Negotiations with the USSR were extended to gas.

By 1966, the new organization of the group had emerged, in all but formal fashion.[12] The company was to be reorganized under four holding companies: AGIP for oil production and marketing, ANIC for refining and petrochemicals, SNAM for pipelines and gaslines, and SNAM PROGETTI for engineering design and research and development. There were separate companies for Lanerossi (textiles), NUOVO PIGNONE (engineering), and AGIP NUCLEARE (nuclear fuel). Cefis insisted on handling over his powers to an executive committee and his general manager, Raffael Girotti, in order to end the one-man rule. A year later, the Italian government repealed the law preventing ENI from entering private ventures, and ENI immediately entered into new agreements with Phillips for exploration in the North Sea and Nigeria. The credibility that these moves gave ENI, together with the acceptance by the oil world that ENI was there to stay, enabled the group to go to the international bond market for a $65 million issue which was oversubscribed.[13] At the same time, ENI was given certain monopoly rights offshore in the Adriatic together with a claim (25 per cent) on the exploration area and first option of purchasing any natural gas discovered by private firms.

1968 was a briefly euphoric year, as it looked as though ENI could finally become self-sufficient in crude, with oil strikes in Libya (a major field), Egypt, Iran, Nigeria, and British and Norwegian sectors of the North Sea, and gas strikes offshore in the Adriatic. ENI hoped that by 1971–72 it could expect the following oil supplies shown in Table 1. In the same year, ENI made an important investment in the chemical sector, buying a strong stake (13 per cent) in Montedison, with a secret open-market operation mounted by Cefis after he failed to get cooperation from the company in the petrochemical field.

The policy of the group's leaders, while still avowedly designed to aid Italian

Table 1. 1968 expected supply sources and refinery capacities, 1971–72[14]

Libya	8–18	mT
Tunisia	3.0	mT
Qatar	1.2	mT
Iran	10–12	mT
Italy	1.6	mT
	23.8–27.8	mT

Refinery capacity, 27.4–30.8 mT.

economic progress, was nevertheless international in its outlook, elaborating Mattei's thinking that oil was too important to be left to oil men, that agreements should be made government to government, and that the host (producer) country should benefit as much as Italy. The primary requirement was to have adequate controlled supplies of raw materials,[15] and between 1969 and 1973 one-third of ENI's capital outlays were made outside Italy. The company began to advertise itself as a multinational enterprise.

In 1969, the company entered the Hungarian retail market and signed a twenty-year gas-supply agreement with the USSR for 6 billion cubic meters per annum. An identical agreement was signed with Holland the following year and production restarted in Nigeria when the civil war ended.

In 1970, the government committed 50 per cent of the autostrada petrol stations that were to be constructed to AGIP, a move that would boost both ENI's market control and profitability.[16] The agreements for the gas pipelines from Holland and the USSR were signed in 1971 for a total of $516 million, and then on April 12, Cefis suddently resigned the presidency to take control of Montedison. The general manager, Raffael Girotti, became president.

Cefis' accomplishments had been notable; from an oil company with no oil and essentially Italian in scope in 1962, by 1971 ENI was within reach of self-sufficiency in oil, and was operating on a truly multinational scale. ENI had 25 per cent of the Italian market, was then the thirty-third largest corporation outside the United States in sales, and was a dominant public corporation in Italy.[17]

Girotti – Political Pressures

The appointment of Girotti as successor to Cefis marked an important departure from ENI's past. For the first time the president was not a man with strong political power who could act, if not independently of the state, at least with something akin to independence. Moreover, the government appointed to the vice-presidency Francesco Forte of the Socialist Party who was intended to introduce a great deal more political control into the running of ENI than had existed to that time. A two-party interest in the management was thereby established as the president was a Christian Democrat.[18] Within ENI, Girotti proved to be an able manager and businessman, but government interference in ENI's affairs had increased.

In 1972, following the announcement of a huge Libyan find capable of 9–14 million tons per annum production, ENI agreed to a 50 per cent participation by the host government. This was entirely consistent with ENI's past policy, and probably prevented the outright nationalization that befell the other Libyan operators who had not been as accommodating. Oil strikes continued to be made in Libya, Iran, the Congo, the North Sea, and the Adriatic. In June, Girotti negotiated a supply agreement with Iraq for 20 million tons of crude over ten years, a move which annoyed the majors since it eased the difficulties they hoped Iraq would encounter as the result of its recent nationalization of their interests. 1972 also saw a trial of strength between Cefis (at Montedison) and Girotti, which culminated in a *modus vivendi* being proposed by CIPE (Interministerial Committee for Economic Planning) between the two companies, making them managerially independent of one another. ENI was to continue in petroleum and basic chemicals and Montedison would deal with fibers and fine chemicals. It was never put into actual practice.

At the end of the year the *de facto* reorganization of ENI achieved under Cefis was announced officially, revealing a structure very similar to that of multinational companies. There were to be four operating divisions and five staff departments. These were, respectively: petroleum division; chemical and nuclear; engineering and services; manufacturing, research, and development; and planning, finance, control, personnel, and foreign state relations. 1972 was notable as the first year in which ENI's own oil activities exceeded its gas activities in energy content.

In 1973, ENI announced an important ($450 million) fiber plant in Sardinia, an investment which seemed to be prompted more by employment considerations than market prospects. In the same year ENI missed an opportunity to expand its market share in Italy when BP sold its Italian operations to an Italian independent oilman, Signor Monti, after first offering it to ENI. Just why ENI missed this opportunity is debatable, but it did not hesitate when a second chain, that of Shell, came up for sale at the end of the year. At a stroke, ENI boosted its share of the total Italian market for oil products from 25 per cent to 33 per cent, its petrol share from 14 per cent to 40 per cent, its refineries by 50 per cent, and its outlets by 60 per cent. The deal became effective on January 1, 1974, and included a crude-supply agreement of 9 million tons for 1974 and 1975, and then 12 million tons spread over the period 1976–78. The Yom Kippur War at the end of 1973 served to underline Italy's crucial dependence on imported oil and the role of ENI in providing Italy with the energy it required, as ENI felt obliged to supply the market at any cost.

Exploration continued particularly in the North Sea, Algeria, Tunisia, Libya, the Congo, Canada, Indonesia, Iran, and Italy. In the middle of 1974, a rich find of gas and distillate was made at Malossa near Milan, a find interesting as much for the techniques of seismic analysis used as for its yield, 2.5 million tons of oil (3 per cent of Italian demand) and 3 billion cubic meters per year (16 per cent of Italian demand).[19] Shipbuilding orders had been placed in 1973 and 1974 that would bring the total fleet size to 2.5 million deadweight tons.

At the end of 1974, Girotti's contract ended and it was not renewed although he stayed on a full year until a successor was found. In the next twelve months, a gas-supply agreement was made with Libya for 3 billion cubic meters per year in exchange for ENI services and goods and participation by Libya in Italian refineries. A similar agreement had been proposed to Iran in 1974, but had not been completed.

Girotti finally resigned in October 1975, and his place was taken by P. Sette, a president of another state holding company in the mechanical sector. Sette, a Bari lawyer, had been a long-time associate of the prominent Christian Democratic leader Aldo Moro.

During Girotti's time in office, ENI had continued to search for secure supplies, but had conceded control of 50 per cent of the Libyan fields, and had lost the Egyptian fields. Price control of the Italian market had been reinforced and investment had continued in Italy for mixed social and market reasons. But, most importantly, this period showed a sharp increase in political interference in ENI's activities, due to both Girotti's weaker political power and ability relative to his predecessors, and to the rise in desire and perhaps also the need of the politicians to control ENI.

Even by the time Girotti took the presidency, various political forces had built up a considerable head of steam, and real political interference in the company affairs had begun with the appointment of Francesco Forte to the vice-presidency. Criticism was leveled against ENI for behaving too much like a private oil company and not taking account of its social objectives as a state concern, for being less than completely transparent administratively and financially, for not putting sufficient effort in searching for hydrocarbons in Italy, and for being a tool of the Christian Democrats (although it had more often been the reverse). The oil scandal covering all the oil companies and the profits they made from the rise in oil prices that was the sequel to the Yom Kippur War also tainted ENI.

Since 1970 (and continuing to the present day), ENI and the other oil companies had been subject to price control, partly for internal domestic reasons and partly because the government felt that oil companies were making immense profits in production which were not being passed on to the consumers, a view to which ENI, as a major arms-length buyer, subscribed.[20] Obviously, as a national company, ENI was in a better position than the multinationals to ride out losses resulting from price control, although ENI, as did the multinationals, greatly increased its indebtedness; this may have been the reason why BP and Shell sold out. Finally, 1973 saw the birth of the Italian Petroleum Plan, published in 1974, which envisaged a larger role for the company, but, taking up ENI's long-term theme of country to country agreements, relegated to ENI only the execution of these agreements.

ENI would no longer be left to work out its own destiny alone, and behind all these changes in ENI's environment was also a power struggle with several other state corporations: Montedison; IRI, the state finance organization involved in nuclear plants supply; and ENEL, the state electricity authority involved in geothermal energy. There is little doubt that Girotti was not fully able to control conflicting pressures, and consequently ENI lost some of the drive that so distinguished it when it was establishing itself.

The arrival of Sette in 1975 marked an interesting point of choice for the future direction of the corporation. The heady period of expansion at all costs under Mattei, where every initiative was worthwhile and questionable decisions were smothered by the speed with which everything was changing, would obviously not occur again. Neither would the freedom with which Mattei and Cefis were able to act, both having an assured cash base and the flexibility and power which go to those who create an empire, rather than those who inherit one.

The short period during which Girotti held the senior post undoubtedly was the beginning of much tighter governmental control and a closer involvement with the aims of the Government that had existed to that time. No politician would happily contemplate controlling an organization with as much unfettered power as ENI had possessed, but with evidence of ENI being involved in a number of Italian oil world scandals, and facing accusations over an accounting system that was at best opaque, the appointment of Sette, the first 'outsider' and a man with experience in running another national enterprise, was the culmination of the efforts to bring ENI under control.

Sette – Assessment and Potential

Coming from another national company, Sette would have been under no illusion that either ENI's or his own performance would be judged on wholly economic criteria. Indeed an oil industry observer sympathetic to ENI suggested in early 1976 that ENI could not be understood unless one understood the national aspirations placed in the company, a point of view easier to understand if one has seen the ENI headquarters in Rome whose appointment and setting goes beyond normal oil industry opulence.

Although the original charter of ENI defined its activities narrowly to 'ventures of national interest in the fields of chemical and nuclear fuels and their related mining', and specific permission was required to initiatives outside these areas, the intention behind the charter was wider than the words suggested and was certainly so interpreted by Mattei and his successors. More or less explicitly other objectives would have included a reduction in dependence on the Anglo-Saxon majors, particularly the Americans, the provision of a direct check on the operations of the multinationals in and outside Italy, a better control of both energy and balance of payments policies, the provision of independent and reliable sources of crude oil, a tool for domestic regional policy and for the acquisition of technology and management expertise, and an instrument of foreign policy, and for all this ENI was also expected to show a return on the government's investment. Sette was well aware of the mixture of political and economic aspirations that he had inherited with the chairmanship, and in measuring the achievements and potential of ENI would have used them to assess his likely impact on the group.

With over 80 per cent of its sales in hydrocarbons ENI was first and foremost an oil company, and in the eyes of the oil industry was subject to two tests – was it able to find oil efficiently and was it able to sell oil efficiently?

In exploration effort, ENI had as much been dogged with bad luck as lack of result. Wanting for muscle in the early days in competing for concessions, ENI was left with areas that were potentially less fruitful and sometimes politically less secure. As a result, despite some other successes, the need to find new sources controlled by the company was still great. To the extent that ENI had not been able to find its own independent oil sources, the company had managed to get round some of the difficulty by intergovernmental agreements. Long-term supply contracts, generally in exchange for goods and services and, consequently, difficult to formulate in terms of 'preferential' or 'beneficial' prices, provided some coverage of the home market – 37 per cent in 1974. However, in terms of independence, the situation was still unsatisfactory as compared to the majors who usually controlled 90 per cent or more of their supplies, whereas ENI only controlled 45 per cent directly in 1974. Having acquired Shell Italiana, ENI was obliged to buy 18 per cent of her home market requirements on the open market in that year; and by 1980, with the same market share, this shortfall would be 21 per cent. It would be worse if, as envisaged in the Petroleum Plan, the market share rose to 40 per cent, when the shortfall would be 24 per cent. Thus, while ENI had achieved some measure of independence, a significant need for open market purchase that was likely to increase in the near future still remained. While the shortage lasted it was raising ENI's crude-supply costs. Such supplies of its own that ENI did possess were reasonably diversified as to country of origin but 72 per cent came from the politically sensitive, oil productive Middle East and although exploration efforts were being directed elsewhere (43 per cent in Asia, 35 per cent in Africa), when compared with the security and diversification of supply that the majors could offer, ENI's capabilities were inferior (see Appendixes 2 and 3 for details and indicative costs).

To transport her crude, ENI had made a considerable fleet investment, and plans were to raise the 1974 fleet of 1.03 million DWT to 2.5 million DWT by 1977.[21] The ultimate intention of the Petroleum Plan was that 50 per cent of the country's demands would be carried in ENI tankers although it was not clear whether the state would provide the necessary finance, particularly when there was a worldwide surplus of tankers for the foreseeable future. In addition to the fleet, ENI owned a pipeline network which carried 20.6 million tons in 1974, and a 10 per cent share in the Trans-Alpine line, which fed the German refinery at Ingolstadt, among others, and carried 25.5 million tons in 1974.[22]

ENI's refineries had only 18.5 per cent of the total Italian refinery capacity but as Italian refineries accounted for 25 per cent of the European Economic Community total,[23] and in view of Italian economic difficulties, there appeared no *a priori* reason for further investment in refining capacity and plans for two refineries (at Portogruaro and Civitavecchia) were shelved in 1975.[24] In view of the country's relative overcapacity of refining, it did not seem likely that ENI would have difficulty in refining sufficient crude to meet her market share, and could continue to use the capacity of other refiners. Indeed in 1976 the capacity of ENI's refineries (40 million tons per year) was equal to 40 per cent of oil demand.[25]

ENI's greatest success had been marketing petrol for, with the acquisition of the Shell network, the company was now by far the largest supplier (40 per cent compared with the next largest, Esso, 15 per cent); and the company had followed an aggressive marketing policy which had benefited the consumer by keeping a downward pressure on prices as well as providing a high standard of service (see Appendix 4). In the field of oil and oil products, the government had both helped and hindered ENI. Onshore and offshore exploration monopolies (in Italy), which had primarily benefited gas production (see below), together with the right to purchase gas from any commercially proven field, had been valuable aids as had been the preferential outlets on the autostrada where ENI would have 50 per cent of these high-volume outlets. Against this assistance, ENI had suffered, along with other companies operating in the Italian market, from product price controls; and ENI's primary obligation to supply the market regardless of cost had been a restriction private companies had not had to face. Hence from the point of view of oil alone on the criteria of the industry, ENI was at best a qualified success. Such success as the group had enjoyed had been dearly bought as the operating losses for 1974 and 1975 had shown (see below). But ENI was more than an oil company, it was also a major supplier of gas, and in 1975 this was still a large part of its activities.

The gas-supply position could not be regarded in quite the same light as oil, as intergovernmental agreement played a stronger role. Present domestic production, accounting for 55 per cent of consumption, was expected to fall to 50 per cent by 1980. ENI had no interests in the production of gas imported into Italy and, indeed, ENI had discovered no significant sources of gas outside Italy. Esso was involved in 38 per cent of expected imports from Libya and Holland; in both cases, supply contracts included a measure of intergovernmental agreement. As no Western European economy had an open market for gas, independence of supply for a company such as ENI did not have a great deal of meaning.

The issue of prices and ENI's gas profits were crucial to the understanding of ENI's relative success, for it was the profit on gas which had paid for the partial and expensive attainment of independent oil supplies, the refining, and marketing. However, Italy (ENI) was not alone in relating gas prices to those of other fuels, regardless of the production price of gas, and in recent years the cost price of gas had changed with the increased expense of offshore gas production and the expense of imports. Also the oil companies, against or instead of whom it was ENI's explicit purpose to react, had financed their activities on the profits made at the production level (by means of transfer pricing), and it was considered acceptable for ENI to have a secure base for self-financing in order for it to achieve its purpose. Nonetheless the extent to which ENI's activities had been cross-subsidized by gas profits was difficult to assess, and estimates of the sum of gas profits varied. In 1974, with prices below the level of fuel oil parity, profits were calculated to be of the order of $470 million,[26] while the total group loss was $89 million. In 1975, Sette declared that the so-called 'methane income' (the net income on natural gas activities) was 170–180 billion lire, but other observers thought this was an underestimate (see Appendix 5). The profits on gas enabled ENI to cross-subsidize activities (including oil) to an extent which was not visible in the accounts. It seemed, however, unlikely that ENI would continue to earn so much on gas because imports projected for 1980 would amount to more than half of consumption (Appendix 5) and profits on these would accrue to the exporter. Thus, the increased activity for the group envisaged by the Petroleum Plan and the decreasing amount of gas profits meant that ENI would have to restructure its internal finances quite radically. The government's attitude to ENI's financing would also have to change, since the previous parsimonious policy, resulting in the low capital base, would not be sufficient in the future.

Historically, the financial policy had deviated only of necessity from Mattei's concept of financing as much as possible internally in order to preserve independence; Cefis used the market for loans for the same reason, seeking government equity only as a last resort. Full use of depreciation had been made (in 1975 the provision nearly doubled to 47.5 billion lire), with large long-term indebtedness, at the same time pressuring the government for equity finance. However, ENI's internal cash generation had decreased from over 70 per cent ten years ago to an expected 40 per cent for 1975 and the 'external' requirements of the group were now considerable. Expected investment for 1974–78 amounted to 5,700 billion lire,[27] and if no further investment was forthcoming from the state, ENI might well have to turn again to the Euromarket, where previously it had had no difficulty in raising loans. However, poor results of 1974 and 1975 and the unstable political situation after the elections in the middle of 1976 made this less likely in the future. Ironically, as ENI was at the threshold of securing an adequate oil supply, domestic gas supplies, which had generated so much cash, were increasingly being replaced by expensive imports. The problem of the future might be the assurance of an adequate finance supply, with the backing of, or from, a government which was all but bankrupt.

Although oil and gas represented the largest part of ENI's activities, Sette could not forget that the group's other businesses represented considerable sales (see Table 2) and several political problems, and any future strategy would need to take them into account.

Included under the sales of chemicals of ANIC were the nuclear fuel sales of AGIP NUCLEARE. Mining and exploring operations were carried out in various parts of the world (Nigeria, Australia, Bolivia, Somalia) for uranium ores, generally under AGIP management, and enrichment and reprocessing agreements were in existence between the USSR and the UK.[29] In addition, ENI had a stake in the European Eurodif Group and conducted a research program into nuclear industrial applications.[30] As a small employer (300 persons), the nuclear fuel activity did not rank highly in ENI's hierarchy of interests. But the increased role for nuclear fuel envisaged in the Energy Plan and criticism of ENEL, the state electricity authority, because it had relied too much on low-cost fuel oil,[31] together with the revelation that oil companies, including ENI, had been financing a campaign to dissuade ENEL from the nuclear option,[32] all combined to make nuclear fuel activity one of great public interest.

Chemical sales were sufficiently large to make ENI the second largest Italian petrochemical group after Montedison. More frequently than any other of ENI's subsidiary companies, chemicals had been used as the vehicle to direct investment

Table 2. Sales outside oil and gas sectors, 1974[28]

Chemicals	$981,500,000
Textiles	222,500,000
Mechanical	99,000,000
Other	53,000,000

to the South; with only 10 per cent of group sales, chemicals provided 27 per cent of group employment and sales in 1974, at $303 million, and represented 24 per cent of total group investment.[33] The products ranged from ammonia and nitric acid, through fertilizers, synthetic rubbers, organic chemicals, and plaster to chemical fibers; and fibers in particular brought ENI into direct conflict with its largest rival, Montedison.

The agreement on spheres of interest arising from the conflict between Cefis and Girotti in 1972 had never really been implemented; competition had continued between the two corporations to the point where both were declaring losses. Suggestions had been made both within and outside the corporations that there should be closer cooperation between them; Montedison had contended the ENI could compete unfairly in the textile sector, having its own captive outlet in Lanerossi, and had suggested that the two companies enter into joint planning agreements for other products. It was a moot point whether or not joint action between these two companies would lead to more efficient investment. Textiles had been a loss ever since ENI had been made to buy Lanerossi (1961), but, nonetheless, accounted for 25 per cent of group employment (i.e. 23,105 persons) while providing only 3 per cent of sales. The exact amount of the losses sustained had never been clear from the accounts, but vice-president Forte had calculated[34] that losses in 1974 were of the order of $16 million and were expected to increase in 1975 to $80 million. The various textile interests of ENI had been grouped under a holder company called TESCON, whose 1974 investments amounted to $40 million (3 per cent of total), almost double the previous year. No amount of rationalization and investment seemed to have had any lasting effect on the textile sector; it was likely that political considerations of the employment generated by TESCON would oblige ENI to continue its investment, although there were also some optimistic reports to the future.[35]

The mechanical company NUOVO PIGNONE provided 1 per cent of sales and 7 per cent of employment; it made various types of compressors and equipment related to petrochemical, nuclear, and textile activities. From the accounts, it was not possible to say whether or not these activities were profitable, and sales from this industrial arm of ENI had more often than not been used as items of exchange in oil-supply agreements.

The sales of two similar ENI companies, SNAM PROGETTI (pipelines, refinery, construction, and design) and SAIPEM (drilling, underseas pipelines, general construction), while sometimes operating independently of ENI's own hydrocarbon sector, were included in the general results of oil and gas operations. The companies that fell into the last category of ENI sales, 'other', were as controversial as any that ENI owned and included, most notably, SOFID, the financing company which was used for all investment purposes and general financing, and *Il Giorno*, one of the largest Italian newspapers with reputed heavy losses, according to vice-president Forte, of $10–15 million in 1974.[36]

Sette was then the inheritor of a number of problems, and he was aware that the rapid growth of ENI over the last two and a half decades was very unlikely to continue (see Appendix 6). It was also clearly expected that he would be under the thumb of the government. However, people once in office are not always as pliable as they appear beforehand, and the basic executive structure of the group gave tremendous power to the president. The Board of Directors, consisting of the president, vice-president, five ministerial representatives, five experts, and two employee representatives, was convened by the president and needed a quorum of nine. The Executive Board, which consisted of the president, vice-president, and three directors, could work with a quorum of three for all 'routine and special administration not specifically reserved to the Board of Directors'.[37] Relations with the vice-president had been problematical but were eased in 1975 when Forte was replaced. He had been in the forefront of a campaign to decrease the opacity of ENI's accounts and organization and had refused to sign the accounts in 1973 and 1974, ostensibly because he felt that the picture presented was entirely misleading,[38] although he may also have had more political motives. Certainly he represented the government's wish to have a stronger say in the running of the group. (See Appendixes 7, 8, and 9 for the balance sheet and income statement for 1975.)

The Politicizing of ENI and the Oil Crisis of 1979

At the end of 1976, when Sette had been in office for two years, no really clear pattern of management had emerged. Fundamental changes had taken place, however, in the political context in which ENI operated.

During the June 1976 elections the Communist Party made serious inroads into the previous Christian Democratic majorities, and the Christian Democrats only remained in power at the tolerance of the Communists. Consequently with the Communist deputies participating in the parliamentary committees, these committees became more significant, whereas before they were essentially a tool of the (Christian Democratic dominated) Cabinet. The Industry Committee had a Socialist chairman and an indirect Communist majority with the help of the left-wing Socialists (six over five). One of the Christian Democratic members of the Committee was a newly elected Senator, Girotti, the former president of ENI. Companies other than ENI – Italian and multinationals alike – were rather pleased with these developments. It was the view of some industry observers that the Communists – and, of course, Girotti – understood the economics of the oil industry better than the Christian Democrats'. It was also probable that the Communists would be critical of ENI because they regarded it as one of the Christian Democrats' domains. Finally, the Communists – so it was suggested – were not saddled with previous governmental policies, particularly those concerning the ceiling prices of petroleum products, and would bring changes to these areas.

Product prices in Italy were set by the government as ceiling prices which, for a variety of reasons, had been kept low. If these prices were not raised, it was estimated by private industry that the industry might lose $1.5 billion in 1976. Because of the profit squeeze some independent companies, notably MONTI, had stopped importing crude and had kept refinery runs at the minimum. If other private companies followed suit, there could be a lack of supply. In this situation, ENI would have to import more to keep the Italian market supplied, thereby increasing its downstream losses. As a matter of fact, in early 1976 the government did tell ENI to increase supplies; Sette asked for these instructions in writing, which the government provided.

The government had been reluctant to raise ceiling prices, which was attributed to either their inability to act or the fact that raising ceiling prices would increase the profits of all the oil companies; this was politically unacceptable both because the private company profits had been a point of attack since the days of Mattei and because more profits would permit more self-financing. The independence given by self-financing had been an object of criticism for politicians of all colors.

According to a report,[39] subsequently perfunctorily denied, Esso informed the government that it would end its Italian operations in early 1977 if crude price increases could not be passed on to the consumer. Esso was the second largest oil company in Italy, with an investment of $1 billion; it had lost $91 million in 1975 and estimated its losses for 1976 at $120 million. Whether or not Esso pulled out would depend on its assessment of the Italian government's intentions. Shell had sold out to ENI in 1974 because it had concluded that the government (and ENI) were malevolent. At that time, Esso decided to stay in Italy but it was possible that the parent concern, Exxon, had now been exhausted, although if Esso could not find a buyer for its facilities – a real likelihood – it would constitute a major write-down for the parent firm. But Esso had 15 per cent of the market and if the government were to direct ENI to take over Esso's operations, that would add considerably to ENI's downstream losses.

To correct these downstream losses – especially in view of the market slowdown in the growth of oil consumption – ENI undertook, beginning in 1977, to rationalize its refining, transportation, and marketing activities. In view of the increased politicizing of ENI, however, actions were slow in coming. Rationalization of operations have been extended to the textile sector. One positive development was the signing of an agreement with Algeria for gas imports. This agreement is not only favorable to ENI but will contribute to one of ENI's goals, namely, the strengthening of the economies of Sicily and Southern Italy.

1978 and the first half of 1979 brought unfavorable financial results to ENI. According to the 1978 Annual Report, the 1978 losses increased by over more than 200 billion lire over 1977. While profits were shown in the oil and gas activities, the above increase in losses was due entirely to the EGAM group. This group, consisting of some twenty-three companies in the mining and metallurgical sectors, six companies in the textile machinery sector, plus some real estate and export–import companies, were, almost literally, forced on ENI by the Minister of State Holdings in mid-1978. While ENI undertook to streamline the activities of the EGAM group, it was clear that ENI was landed with yet another white elephant. In the first half of 1979 – in the aftermath of the OPEC price increases of some 60–65 per cent – ENI's oil and gas profits increased substantially over the first half of 1978, though much of the profits were due to the rise in the value of inventories. Other ENI activities – outside oil and gas – continued poorly. 1979 also demonstrated the exposure of ENI to the vagaries of OPEC supplies and pricing and to the uncertainties of Italian domestic politics.

According to the 1977 Annual Report ENI had to purchase 58 per cent of its crude requirements from other suppliers. Thus the OPEC price increases hit ENI particularly hard. Moreover in 1979 demand increased rapidly as the government was reluctant to increase domestic product prices in line with the crude price increases because of the elections that were held in June. In response to the surging demand and its own inadequate supply, ENI mounted a two-pronged attack. In the short run, ENI was able to conclude new supply contracts with Iraq, Iran, and Saudi Arabia thus covering ENI's immediate requirements.[40] For the long run, ENI planned to invest about $17 billion[41] over the next five years – an increase of 40 per cent over the previous five-year period – mostly in exploration for oil and gas.

1979 also witnessed a change in the political environment in which ENI operated. Hitherto ENI was considered the domain of the Christian Democratic Party. This changed as a result of a coalition package affecting the three major state holdings. According to this agreement, the ENI chairmanship went to the Socialist Party and Giorgio Mazzanti, 51 and a chemist by training, replaced Sette on February 1. Mazzanti's tenure lasted but ten months as in early December he was suspended from his position because of allegations that huge commissions were paid to Saudi and Italian officials as part of the Saudi Arabian–ENI deal. Also in early December, the Saudi government suspended some of the contracted crude shipments to Italy.[42] The executive situation further deteriorated in early 1980. Although Mazzanti was cleared of any wrongdoing, he resigned his post, and the government appointed Egidio Egidi to the chairmanship. However, Egidi resigned one day before he was scheduled to assume the position. Egidi served as the government's special commissioner following the suspension of Mazzanti. Egidi allegedly resigned because of increasing governmental interference into ENI's affairs.[43] Thus the chaotic Italian political situation finally was transferred, in full, onto ENI.

In view of the supply and executive uncertainties the outlook for ENI was variable. If profitability had been the sole objective, the creation of ENI would not have been undertaken. Indeed although comparisons across countries with different market characteristics and different economic conditions should be subject to caution, in comparison with other oil companies of a similar size ENI was notably unprofitable (see Appendixes 10 and 11). The low profitability of ENI could have been accounted for by the need of a national company to fulfill various roles which, while uneconomical, were deemed to be 'in the national interest' and, in the case of ENI, the dominant 'national interest' had been the creation of employment (principally outside the hydrocarbon sector). Thus, management tolerated a level of staffing which made oil production and refined products produced per employee by AGIP, for example, low by comparison with the other companies. Sales per employee for AGIP however were comparable with those of other companies. The low profitability of ENI could also have had a more obtuse explanation, in that ENI's principal shareholder, the state, could probably only have been induced to supply additional capital for investment if the group had demonstrated the need for it with a low profit or, better still, a loss. And the group's shortage of capital was reflected by the debt structure which, while not unusual by general European standards, was high in comparison to other oil companies (see Appendix 12). Without more detailed accounts there was little evidence to suggest that the various subsidiary companies (see Appendix 13 for the 1976 organization) of ENI were not efficiently managed, barring the effects of government interference. And insomuch as it was by far the largest company in the country and that Italian big business was highly politicized, ENI was as much a subject of government interference as ENI interfered with the affairs of the government. But if the executive potential existed for an efficiently managed company, the financial independence looked less assured. ENI's original source of strength – the profits made on gas – seemed to have a less lucrative future, with an increasing limitation of domestic gas, price control, and demands for increased visibility of the cross-subsidizing than had occurred in the past. In addition to this, the financial performance of ENI in recent years made it a poor prospect for loans either from the Italian banking system, where it had favored access, or on the international money market, a problem compounded by the fact that its guarantor and alternative fund source, the government, was not only nearly bankrupt, but also politically paralyzed. At the end of 1976, ENI had accumulated debts of 4,000 billion lire.

On the technical level, the picture was more encouraging. For its hydrocarbon activity the group possessed acknowledged skills at all levels and in all fields which made it comparable with any other major oil company. Indeed, in some fields, notably underseas pipelaying, ENI was a world leader. In engineering, too, particularly in relation to hydrocarbon work, the group's products met with respect from its competitors. With its expertise and experience, ENI was now able to play a role as the state energy corporation. To be sure, its independence of oil supply did not correspond to that of the multinational groups, but it was sufficient for the group to be a potent 'other force' although possibly still not enough to provide adequate emergency security. But the objective of reducing Italy's dependence on the 'Seven Sisters' and providing Italy with a secure and inexpensive supply of oil had not been reached. ENI's inability (or bad luck) to find sufficient crude oil of its own meant that it had to buy it on the open market at commercial prices from the oil multinationals. With the assertion of OPEC power, beginning in the early seventies, it was no longer clear, however, what 'secure' supplies meant, even to the majors.

In chemicals and textiles (both large employers), the situation was considerably less assured, and the latter particularly appeared to present for ENI insoluble problems, being too large to disband and too big to ignore. The germs of ENI's future problems lay perhaps in the area of nuclear fuel with clashes of interests with other state corporations operating in the same related fields, particularly ENEL, but also with IRI, the state financing group, and with Montedison, in which the state had an indirect stake. The complex interplay between ENI management and the politicians undoubtedly had led to some economically dubious ventures and increased political interference might even augment this.

Nevertheless, ENI had a remarkable record of achievements, particularly if one considered that it was 27 years old as of 1980. ENI was the eleventh largest oil company, in the world – the third largest national oil company and the twenty-second largest industrial corporation in the world (*Fortune*, August 1976). The government had successfully used ENI to bolster the economy of the South, to acquire managerial and technological expertise, and, importantly, to enhance Italy's prestige both at home and abroad. In a country which the rest of Europe viewed with some amusement and where, which was worse, many Italians considered themselves with a national inferiority complex, ENI's achievements could reasonably be viewed with a sense of pride.

The Italian Energy Context 1975–80

Of the industrial countries, Italy was second only to Japan in its dependence on external fuel supplies, and was unusually dependent on oil, the dominant fuel from the start. Italy's rapid growth in the fifties and sixties necessitated an expansion of the oil supply (see Appendix 14). However, continuing ability to supply this demand was limited. Italy's oil reserves were negligible and her gas reserves amounted to only 1 per cent of world reserves (consumption was 1.5 per cent of the world total in 1974[44]). Finally, there was no coal.

The Energy Plan announced in 1975 sought to redress the imbalance in Italy's energy make-up and to provide better coordination between the various state companies involved in energy production, of which the two most important were ENI and ENEL. It was also a purpose of the plan to make the country less dependent on oil supplies by developing alternative sources of energy, and to this end the plan opted heavily for nuclear power. Besides the issue as to whether a real independence was achieved with this option, it was an open question as to whether Italy had the capability to bring so much nuclear power into action that quickly, given the multiplicity of designs available, the long lead times, and the relative novelty of the art. Plant costs also were high and local opposition forces strong.

Regardless of how these arguments would be resolved, Italy would depend on oil and gas for over three-quarters of its energy in 1985. With 80 per cent of ENI's present sales in hydrocarbons, the plan intimately involved ENI and the Italian government's view of its future role. In addition, as supplier of nuclear fuel, it was closely involved in the nuclear initiative. It was proposed that ENI should control 40 per cent of the Italian oil products market; the merger with Shell has given ENI 33 per cent to date. ENI was to continue to be given preferential petrol outlets, and the total number of outlets in Italy was to be reduced, which would benefit ENI as well; with 11,300 outlets out of 38,000[45] (28 per cent), and 40 per cent of the petrol market, ENI was already 'rationalized'. According to the plan, the refinery situation was also to be improved with new investments near consumption (the North) rather than near unemployment centers (the South), although at present there existed considerable overcapacity. ENI was given the responsibility for ensuring the country's emergency supplies at whatever cost, for which it would be reimbursed by the state, though it was not clear how.

The plan stipulated that pricing was to be in accord with actual costs. Primarily directed toward petroleum costs, which the government had held down in recent years to the loss of every company marketing in Italy, it would alleviate market distortions. But the government had also been holding gas prices down; unless those prices were increased to align with those of fuel oil, ENI would lose the source of her long-term strength, as well as having to adjust to a different energy balance.

As reported in ENI's 1977 Annual Report, in that year the 1975 Energy Plan was revised in several respects. The emphasis of the 1977 plan was on conservation. Because of the higher prices – though product prices were still controlled by the government – the rate of increase of oil consumption was slower than it was assumed to be in the 1975 plan for 1985. Assuming further real price increases, the revised plan called both for incremental conservation efforts as well as the development of new sources of energy to reduce the projected import requirements for oil and gas by 1985.

Appendix 1. ENI's Initial Organization and Capital

Initial organization (nominal)

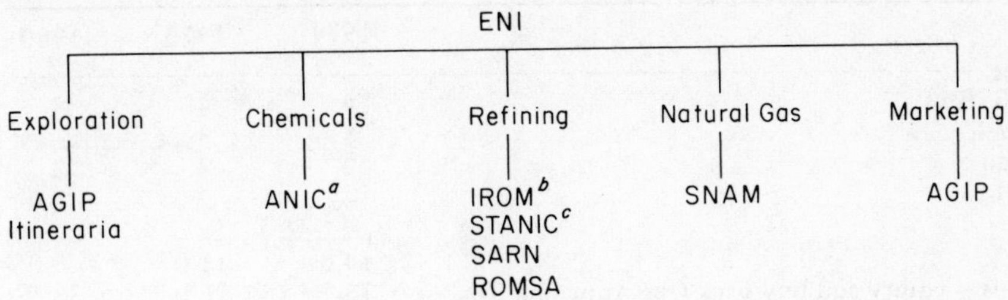

Notes [a]ANIC – 25% AGIP, 25% IRI, 50% Montecatini
[b]IROM – 51 % AGIP, 49% Anglo Iranian
[c]STANIC – 50% ANIC, 50% Esso

Initial assets

	Millions of lire	Millions of dollars[a]
State holdings in AGIP, ANIC, ROMSA, SNAM, ENM	4,582	7.33
ANIC debentures	1,600	2.56
Premises at 43 via Lombardia, Rome	218	0.35
Assessment of income from petroleum exploration in Italy on behalf of state	8,600	12.76
	15,000	24.00

[a]1 lira = $0.0016.

Appendix 2. Oil Supply and Requirements

Crude-supply agreements, by ENI[46] (millions of tons)

		1974	1977	1980
USSR		4	4	4
Shell		8	3	–
Iraq		2	2	2
Libya		–	2.0	2.0
		14.0	11.0	8.0
Plus – equity and buy-back (see Appendix 3)		15.7	26.0	33.5
		29.7	37.0	41.5
Plus – Italian production		1.1	2.0	3.0
'Own' supply	(A)	30.8	39.0	44.5
Italian market requirements	(B)	37.4	44.5	56.1[a]
Refineries overseas		7.0	10.5	11.5
Total market requirements	(C)	44.4	55.0	67.6
Italian arms-length purchase	(B–A)	6.6	5.5	11.6
Total arms-length purchase	(C–A)	13.6	16.0	23.1
Percentage of Italian market requirements	(B–A/B)	17.6	12.4	20.7
Percentage of total market requirements	(C–A/C)	30.6	29.0	34.2

[a]Assumed 34% of market (as in 1974) and market as per Italy Petroleum Plan.[47]

Appendix 3. Oil Supply and Costs

Supply

Equity and buy-back production (in addition to supply agreement noted in Appendix 2)

	1974	1975	1977	1980
Libya	3.9	4.8	7.0	10.0
Tunisia	1.0	0.6	2.0	2.0
Iran	4.0	1.6	5.5	6.5
Qatar	2.5	1.9	3.0	3.0
Nigeria	3.3	2.9	4.0	4.0
Congo	0.8	0.6	1.5	2.0
North Sea	0.2	1.0	3.0	6.0
	15.7	13.4	26.0	33.5

Reserves (equity oil shares) 287 million tons. (Total consolidated reserves discovered: 521.4 mT).

Costs

Cost in dollars per ton c.i.f.

	1973	1974[48]
AGIP	30.5	82.5 (including Shell)
Esso	27.6	74.4

AGIP + Shell	AGIP without Shell	MNC	Average[49]
87.0	89.3	85.8	86.0 for 1975

Note: These costs are indicative only, and do not allow for differences in crude quality; the prices for the MNC are obviously transfer prices.

Appendix 4. Refining Capacity and Product Market Share[50]

Refining

78 per cent sent for refining in Italy
9 per cent sent to overseas affiliates
13 per cent arms length sales, stock variation, chemical feedstock

Refinieries[51] (thousands of tons)

8 Italian refineries. Total refined	28,190
6 Overseas refineries. Total refined (Germany, Tunisia, Ghana, Zaire, Tanzania, Zambia)	7,007
	35,197

Market share total[52] (per cent)

	AGIP	IIP (formerly Shell)	Total (ENI)	1974
Petrol	26.6	13.3	39.9	
Diesel	25.3	10.2	35.5	
Tractor diesel	33.7	9.1	42.8	Total of
Heating oil	22.0	9.7	31.7	all products
Fuel oil	21.1	8.7	29.8	34 per cent
Lubricating oil	16.3	11.4	27.7	
LPG	21.7	13.1	34.8	

87 per cent of products sold in Italy[53]
13 per cent of products sold by overseas affiliates

Petrol stations

	In Europe	1,280 (i.e. outside Italy)
	Africa	888
	Italy	7,141
Plus, from	Shell	4,200

Appendix 5. Gas Production and Supply[54]

Supply (millions of cubic meters)

	1974	1975
Italy		
Group production	14,060	13,420
from underground stock	1,071	600
from third parties	728	620
Overseas		
Libya	1,777	2,245
Holland	1,624	4,164
USSR	747	2,288
	20,007	23,337

Demand (per cent)[55]

Industrial	5	Electricity	6
Chemicals	9	Domestic	33

Predicted imports[56] (billions of cubic meters)

	1974	1977	1980
Libya	1.8	3.0	3.0
Holland	1.6	6.0	6.0
USSR	0.7	6.0	7.0
Algeria	–	–	2.0
Total imports	4.1	15.0	18.0
Domestic production	15.3	15.0	17.0
	19.4	30.0	35.0

Estimated profit, 1975[57]

Sales	21.5 billion m^3
Sales price	\$0.0435 per m^3
Cost price	\$0.0153 per m^3
Profit	\$606 million

Appendix 6. The Growth of ENI, 1974–75[58]

	1954	1964[a]	1971[b]	1974[c]	1975 (b)
			Sales by sector (millions of US $)		
Oil, gas, engineering, and construction	291	982	2,633	7,687	
Chemical	11	116	294	919	
Mechanical	3	42	78	99	NA (b)
Textiles	–	67	165	223	
Miscellaneous	–	11	37	53	
	305	1,218	3,207	8,991	
Operating					
Gas production (million m³)	2.7	7.3513	12.399[13]	14.060	13.420
Gas pipeline (km)	2,803	5,005	9,714	11,911	12,634
Crude oil production (thousand tons)					
Italy	124	2,732	1,314	1,069	1,029
Overseas	–	6,130	10,462	13,767	13,370
Fleet deadweight	100,000	467,764	524,398	1,170,378	1,370,313
Refined products (thousand metric tons)					
Italy	4,932	13,175	19,356	28,190	24,248
Overseas	–	2,370	7,513	7,007	7,434
Fertilizer (thousand metric tons)	–	271	404	502	515
Synthetic rubber	–	110	141	162	126
Synthetic resin	–	64	222	342	316
Chemical fibers	–	–	37	48	80
Employees (thousands)	16	59	76	92	100

[a] 1 lira = $0.0016 for 1964.
[b] 1 lira = $0.0017 for 1971.
[c] 1 lira = $0.0015 for 1974.

Appendix 7. Balance Sheet (Italian Lire)

ASSETS

	1975 (b)	1974 (b)
Property, plant, and equipment		
Nonindustrial buildings, land	10,146,024,182	7,106,663,399
Office furniture and equipment	919,268,099	526,216,399
Library	168,768,458	140,553,893
Property, plant, and equipment in course of acquisition	402,111,400	394,111,400
Deferred charges		
Discounts on bond issues	47,540,806,344	27,388,753,375
Other deferred charges	9,688,399,085	
Financial investments and credits		
Shareholdings	691,495,890,551	721,851,576,795
Loans to Wneshtorgbank for the gas pipeline from Russia	49,263,546,042	57,573,059,832
Loans to subsidiaries and affiliates	1,405,771,082,380	1,402,889,306,459
Advances for subsidiaries' share capital increase	14,396,339,217	
Government investment capital: amount not yet paid in	63,000,000,000	101,000,000,000
Accounts receivable		
Suppliers	8,167,721	6,945,533
Earnings	688,364,227	385,451,911
Subsidiaries and affiliates	107,368,184,696	32,176,218,382
Balance due on bond issues	230,852,965	
Others	8,738,869,307	9,582,838,049
Cash guarantee deposits outstanding	12,963,390	12,343,390
Bonds		
Issued by ENI and redeemed	46,727,250,000	
Issued by others	12,184,631,200	6,772,719,424
Banks and postal accounts	10,034,974,563	7,732,885,358
Active rates and credit receipts (with interest)	17,104,186,719	2,722,582,327
Current profits	7,661,020	675,000
Previous years' losses	53,794,682,122	54,225,828,335

(*Continued*)

Appendix 7 (*continued*)

Net loss for the year	87,112,395,115	–429,311,001
Contra accounts		
Guarantees issued by third parties	44,619,000,000	6,276,000,000
Third parties for guarantees issued, third parties for guarantee bonds, third parties for swaps on securities	854,314,890,842	297,280,588,908
Other contra accounts	773,408,593,750	926,278,655,940
Total	4,309,597,903,395	3,661,894,693,108

Executive for Administration and Finance
LEONARDO di DONNA

For the Board
PIETRO SETTE, Chairman

Appendix 8. Balance Sheet (Italian Lire)

LIABILITIES

	1975	1974	(*b*)
Capital and reserves			
Government investment capital authorized	1,106,900,000,000	1,090,900,000,000	
Statutory reserve	10,149,154,914	10,149,154,914	
Monetary adjustment reserve (Ex Ente Nazionale Metano)	396,405,764	396,405,764	
Special reserve from surpluses on shareholdings	4,779,555,813	4,779,555,813	
Monetary adjustment reserve (law 576–12.2.75)	1,625,214,940	–	
Depreciation			
Nonindustrial buildings, land	4,910,178,814	3,457,296,008	
Office furniture and equipment	468,472,485	302,461,180	
Library	91,874,837	75,811,038	
Employee severance and retirement provisions	9,723,242,036	7,402,779,860	
Deferred tax provision	326,693,739	9,105,841,208	
Sundry provisions			
Shareholding devaluation	494,999,999	–	
Doubtful debts	14,396,339,217	–	
Studies and research	1,186,120,135	1,210,044,595	
Various risks and miscellaneous	21,672,800,000	4,603,375,000	
Deferred income	1,818, 194,441	3,053,041,065	
Financial indebtedness			
Bond loans	973,293,875,000	759,467,750,000	
Loans secured by assets	19,258,201,933	19,640,061,000	
Bank loans	153,000,000,000	93,000,000,000	
Reserve ıeports	–	28,500,000,000	
Accounts payable			
Suppliers	350,682,604	314,981,404	
Earnings	–	130,663,403	
Subsidiaries and affiliates	7,205,884,663	8,330,558,067	
Bonds drawn for redemption	2,869,664,000	2,868,097,000	
Coupons and other debts payable to bondholders	274,143,037	189,659,979	
Others	21,135,450,900	14,064,947,680	
Bank overdrafts	246,518,163,307	309,532,533,696	
Reserve bills of exchange	–	35,000,000,000	
Guarantee deposits in cash by third parties	8,625,000	8,625,000	

(*Continued*)

Appendix 8 (*Continued*)

Accruals and prepaid income	34,400,231,225	25,501,560,236
Frozen assets	1,250,000	74,244,244
	2,637,255,418,803	2,432,059,448,260
Contra accounts		
Third parties for guarantees issued	44,619,000,000	6,276,000,000
Guarantees issued to third parties	854,314,890,842	297,280,588,908
Other contra accounts	773,408,593,750	926,278,655,940
Total	4,309,597,903,395	3,661,894,693,108

The Auditors
ANTONINO CAVALLARO, Chairman, FAUSTO ABATE, LUIGI ACROSSO, ANTONIO CERONI, SALVATORE PAOLUCCI

Appendix 9. Profit and Loss Account (Italian Lire)

	1975
COSTS AND OTHER CHARGES	
Purchases of materials	
For consumption	136,297,979
For investment	554,263,409
Employment cost	
Salaries, wages, and contributions	13,428,869,570
Allocation to employee severance and retirement provision	2,619,424,904
Other costs	1,425,873,499
Services	
Financial	155,943,298
Sundry	19,718,621,673
Sundry costs	723,210,180
Interest and other financial charges	
Bond loans	66,323,321,084
Bank loans	35,625,534,204
Loans from other affiliates	5,766,955
Other debts	2,550,043
Premiums on swaps	3,616,630,642
Other financial charges	27,715,278
Depreciation and amortization	
Property, plant, and equipment	275,265,909
Deferred charges:	
Discounts and other charges on bond issues	9,979,292,836
Other deferred charges	50,144,000
Taxes	
Current year	4,698,644
Previous years	46,429,150
Sundry and extraordinary charges	
Provisions:	
Shareholding devaluation	15,494,999,999
Doubtful debts	14,396,339,217
Various risks and miscellaneous	2,392,425,000
Devaluations:	
Bonds	1,563,008,224
Shareholdings	33,868,418,744
Loss on shareholdings, exchange, disposal of assets, and other charges	64,499,508,200
Total	286,934,632,641

	1975
Revenues	
Services	9,799,681,228
Others	16,360,045,958
Capitalized expenditure	
Materials for investment	554,263,409
Interest and other financial revenues	
Dividends from:	
Subsidiaries	4,979,611,320
Affiliates	3,757,228,692
Other shareholdings	1,125,328,000
Interest on:	
Bonds	3,712,450,149
Bank and postal accounts	1,197,131,323
Credits to subsidiaries	151,742,439,038
Other credits	4,597,135,715
Sundry financial revenues	808,927,951
Sundry and extraordinary income	
Real estate investment	433,959,000
Other sundry and extraordinary income	754,035,743
Loss for the year	87,112,395,115
Total	186,934,632,641

Executive for Administration and Finance
LEONARDO di DONNA

For the Board
PIETRO SETTE, Chairman

The Auditors
ANTONINO CAVALLARO, Chairman
FAUSTO ABATE, LUIGI ACROSSO
ANTONIO CERIONI
SALVATORE PAOLUCCI

Appendix 10. Company Comparisons[59] – Operating Figures

		Standard Oil Indiana	Shell Oil USA	Phillips	CFP	AGIP	ENI
Gross income (turnover)	(million $)	10,025	8,418	5,106	9,120[h]	6,139[c]	8,748
Net income	(million $)	970	620	430	373[h]	64	(89)
Assets	(million $)	8,915	6,129	4,028	7,818	4,175	5,493
Long- and medium-term debt	(million $)	1,427	976	658	874	1,014	1,351
Yearly investment	(million $)	1,826	929	628	234	905	1,274
Issued capital	(million $)	1,875	233	191	227	450	1,636
Employees		47,217	32,287	30,802	27,800	35,875[a]	92,420
Production: oil (million tons per year)		45.6	30.6	11.2	70.0	16.7[b]	16.7
gas (million m^3 per year)		39,392	21,802	7.67[g]	–	15,300	15,300
Refined products (million tons per year)		61.6	55.3	34.4	58[f]	37.4	37.4

Notes: 1 lira = $0.0015, 1 FF = $0.213 } October 31, 1974

[a]AGIP 1974 report gives 7,919 employees. ENI report gives 28% of 78,535 employees in gas or oil sectors, to which are added 13,855 overseas employees, assumed to be in gas or oil sectors.

[b]Equity and buy-back. Total production was 41.2 million tons.

[c]ENI report gives sales of $7,488 for oil, gas, engineering, and construction.

[d]Fondo di dotazone (establishment fund).

[e]Profit of holding company ENI was $0.64 million. This figure quoted by DAFSA.

[f]Sales at home and abroad.

[g]Million tons per year.

[h]Consolidated results.

Appendix 11. Company Comparisons – Ratios

	Standard Oil Indiana	Shell Oil USA	Phillips	CFP	AGIP	ENI
Profit/sales (per cent)	9.7	7.4	8.4	4.1	1.0	(1.0)
Profit/assets (per cent)	10.8	10.1	10.7	4.8	1.5	(1.6)
Sales/employee (thousands of $)	212	260	166	328	171[a]	95
Assets/employee (thousands of $)	189	190	131	281	116	59
Yearly investment/employee (thousands of $)	39	29	20	8	25	14
L & M debt/total liabilities (per cent)	16	16	16	11	24	25
Oil production/employee (thousand tons per year)	0.97	0.95	0.36	2.52	0.47[b]	0.18
Refined products/employee (thousand tons per year)	1.30	1.71	1.12	2.08	1.04[c]	0.40

[a] If sales figures for hydrocarbon sector given in ENI report is taken ($7,488 million), then sales/employee = $208,000. If AGIP report figure of employees is taken (7,919), the figure is $775,000; if overseas employees are added to this (13,855), the figure is $281,000.

[b] If AGIP report figure of employees is taken (7,919), production/employee is 2.11 thousand tons; if overseas employees are added to this (13,855), the production/employee is 0.76 thousand tons.

[c] If AGIP report figure of employees is taken (7,919), refined products/employee is 4.72 thousand tons; if overseas employees are added to this (13,855), refined products/employee is 1.71 thousand tons.

Appendix 12

Capital structure

		Millions of lire
Establishment fund[60]	1953	30,000
Reinvested state profits*	1954	2,300
	1955	2,300
	1956	2,300
	1962	36,900
	1964	125,000
	1966	150,000
	1968	256,000
	1969	211,000
	March 22, 1971	10,000
	July 28, 1971	290,000
	February 1, 1974	12,000
	October 22, 1974	4,000
	August 3, 1975	16,000
		1,110,900

Note: 1975 report of ENI gives total of 1,106,900.
*Annual profits assigned: 20% ordinary reserve fund
15% scientific and technical research in hydrocarbon field and training
60% to state

Consolidated financial structure, 1974[61]

	Billions of lire	Per cent
Short-term bank loans	1,378.1	28
Medium- and long-term loans	2,404.0	49
Endowment fund and reserves	1,132.5	23
	4,914.6 billion lire	

Appendix 13. ENI Subsidiary Companies 1976

AGIP	Oil, gas, and uranium exploration and production, refining and distribution
SNAM	Natural gas transmission and distribution, transportation of oil and petroleum products
ANIC	Chemical activities
AGIP NUCLEARE	Nuclear fuel cycle activities
SNAM PROGETTI	Engineering and contracting
SAIPEM	Construction related to hydrocarbons activities
NUOVO PIGNONE	Mechanical manufacturing activities
TESCON	Textile activities
SOFID	Financing

Appendix 14. Italian Energy Consumption

Sources[62]	1955 (per cent)	1975 (per cent)
Coal	25	8
Gas	6	15
Oil	38	70
Hydro and geothermal	31	7
Total energy	100	100
Total consumption (million tons oil equivalent)	40	138

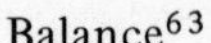

Balance[63]

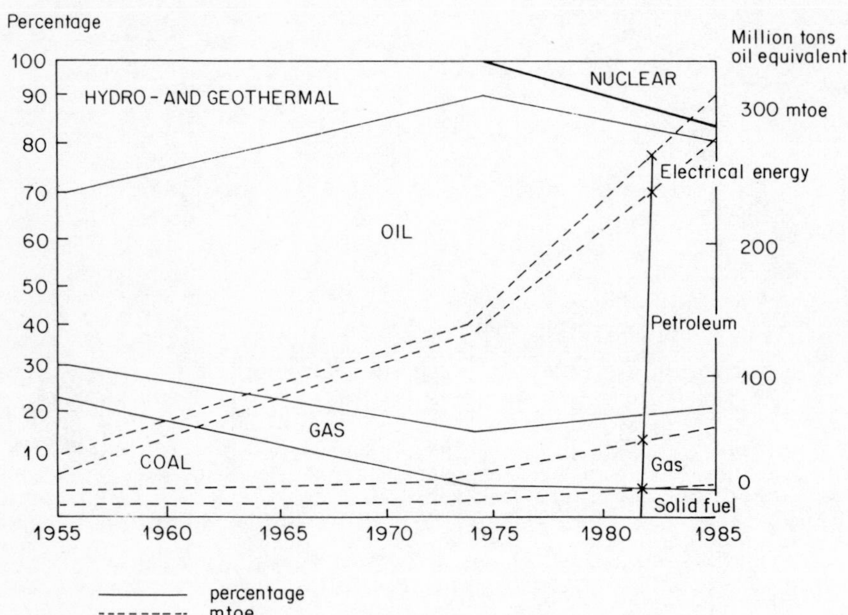

Appendix 15. Ente Nationale Idrocarburi (Dates)

1926	AGIP created.
1936	Iraq concession (BOD Ltd.) sold.
1945	Mattei appointed Commissioner of AGIP, Northern Italy.
1946	Gas discovered at Caviaga.
1949	La bomba petrolifera di Cortemaggiore.
1950	Entire Po Valley resources assigned to AGIP.
1951	AGIP not admitted to Iran consortium.
1953 Feb.	ENI created, Mattei appointed Chairman.
1956	Acquisition of NUOVO PIGNONE Engineering Oil Equipment.
1956	Creation of SNAM PROGETTI, SOFID, AGIP NUCLEARE.
1957	Mining law preventing joint ventures with private enterprise.
	Joint venture agreements with Iran and Egypt.
1959	Petrol prices war started in Italy. ENI reduces pump prices by 25 per cent.
1961	Acquisition of 50 per cent of Lanerossi, textiles.
	USSR supply agreement.
1962 Oct.	Enrico Mattei's death.
1962	Boldrini created President, Cefis created Vice-President.
1963	Five-year supply agreement with Esso.
	USSR supply agreement renewed for six years.
1964	Acquisition of Ragusa field and five-year supply agreement with Gulf.
	AGIP joins Phillips exploration consortium.
1965	Nuclear fuel element agreement with UK Atomic Energy Authority.
	Gas-supply agreement with Esso for Libyan LNG.
1967	Cefis nominated President.
	ENI allowed to enter joint ventures.
	$65 million international bond issue.
	Special status given to ENI for Adriatic offshore.
1968	Oil strike in Libya, Egypt, Nigeria, Iran, and North Sea.
	Gas strike in North Sea and Adriatic.
	Interest acquired in Montedison.
1969	Twenty-year gas-supply agreement with USSR.
	Enter Hungarian market.
1970	Twenty-year gas-supply agreement with Holland.
1971	USSR and Holland gas pipelines agreements.
	Cefis resigns, Girotti appointed President, Forte appointed vice-president.
	Algerian petrochemical and Libyan refinery contracts.
1972	Agreement to 50 per cent Libyan participation for large field.
	Iraq supply agreement.
	Accord on discord between Montedison and ENI.
1973	Acquisition of Shell assets.
	Supply agreement.
	Yom Kippur war.
1974	Malossa oil and gas strike.
	Italian Petroleum Plan.
1975	Gas-supply contract with Algeria. Crude-supply contract with Libya.
	Girotti resigns. Sette appointed President.

References

1. 'The ENI Group's Main Features', Report to OECD by ENI, Paris, March 14, 1962.
2. P. H. Frankel, *Mattei Oil and Power Politics*, Faber and Faber, London, 1966.
3. 'ENI after Mattei', *Management Today*, October 1966.
4. ENI–CEI case, October 1973.
5. 'ENI after Mattei', *Management Today*, October 1966.
6. 'ENI after Mattei', *Management Today*, October 1966.
7. ENI–CEI case, October 1973.
8. Private communication to author, April 1976.
9. ENI–CEI case, October 1973.
10. Private communication to author, April 1976.
11. Private communication to author, April 1976.
12. 'ENI after Mattei', *Management Today*, October 1966.
13. ENI–CEI case, October 1973.
14. 'AGIP Looks to Its Own Oil Surplus', *The Times*, November 28, 1969.
15. ENI–CEI case, October 1973.
16. Private communication to author, May 1976.
17. 'ENI after Mattei', *Management Today*, October 1966.
18. P. Tumiati, 'ENI at the Heart of the Oil Storm', *Financial Times*, October 16, 1973.
19. 'Italy's Promising New Field', *The Petroleum Economist*, February 1975.
20. P. Tumiati, 'ENI at the Heart of the Oil Storm', *Financial Times*, October 16, 1973.
21. ENI – Relazione e Bilancio al Deciembre 31, 1974.
22. Private communication to author, 1975.
23. *Energia ed Idrocarburi: Sommario Statistico, 1955–1974*, ENI publication.
24. Private communication to author, 1975.
25. Private communication to author, March 1977.
26. Private communication to author, 1975.
27. Private communication to author, 1975.
28. *Annual Report 1974*. English version.
29. ENI – Relazione e Bilancio al Deciembre 31, 1974.
30. ENI–CEI case, October 1973.
31. *Key Trends in Countries with an Aggressive Policy*, OECD publication.
32. G. Hodgson, 'Pay-off for the Oil Companies Hand Outs: £500 million', *Sunday Times*, April 18, 1976.
33. ENI – Relazione e Bilancio al Deciembre 31, 1974.
34. R. N. McInnes, 'Italian Watergate?', *Barrons*, 1975.
35. 'Italy's Bright News is Economic', *The Economist*, May 15, 1976.
36. R. N. McInnes, 'Italian Watergate?', *Barrons*, 1975.
37. *ENI Charter of Incorporation and Regulations*, Law 136 of February 10, 1953. ENI publication, 1968.
38. R. N. McInnes, 'Italian Watergate?', *Barrons*, 1975.
39. *International Herald Tribune*, November 20–21, 1976, p. 9.
40. *New York Times*, September 20, 1979.
41. *Wall Street Journal*, July 24, 1979.
42. Detailed accounts appeared in the *Wall Street Journal*, December 6 and 11, 1979, and the *New York Times*, December 6, 1979, and April 16, 1980.
43. *Wall Street Journal*, May 1, 1980.
44. *Energia et Idrocarburi: Sommario Statistico, 1955–1974*, ENI publication.
45. Private communication to author, April 1975.
46. Private communications to author, April 1975.
47. *Italy – Petroleum Plan* (translation).
48. Private communication to author, May 1975.
49. Private communication to author, April 1976.

50. ENI – Relazione e Bilancio al Deciembre 31, 1974.
51. Private communication to author, April 1975.
52. Private communication to author, April 1975.
53. *AGIP – Annual Report and Statement of Accounts*, December 31, 1974.
54. ENI – Relazione e Bilancio al Deciembre 31, 1974.
55. *Energia et Idrocarburi: Sommario Statistico, 1955–1974*, ENI publication.
56. Private communication to author, April 1975.
57. Private communication to author, April 1975.
58. ENI – Relazione e Bilancio al Deciembre 31, 1974.
59. *DAFSA Informations*, 1975.
60. *ENI Charter of Incorporation and Regulations*, Law 136 of February 10, 1953. ENI publication, 1968.
61. Private communication to author, 1975.
62. *Energia et Idrocarburi: Sommario Statistico, 1955–1974*, ENI publication.
63. *Italy – Petroleum Plan* (translation).

CHAPTER 6

Veba: German National Oil Company

Through the sixties and seventies the German economy had been the strongest in Europe and one of the strongest in the world. German economic management was the focus of admiration for many outside observers not least for the articulate belief of the government in the 'social market economy' of Ludwig Erhard and the 'Geordnete Markt' (orderly market). In the field of petroleum products (as in many others) the free interplay of market forces had generally been allowed to determine supply and demand, but concern with security of supply had prompted the government to encourage the formation of a German exploration company, Deminex. The increasingly unstable oil markets in 1972 and 1973 persuaded the government to go further and foster, through merger, the creation of a national oil company, Veba, in which it would have a substantial stake which would give it a controlling interest in Deminex. The merger (between Veba AG and Gelsenberg AG, both companies with principal interests in electricity generation, petroleum products, chemicals, and trading) had been mooted before but never acted on, and the government may have been reacting to past events; in 1966 the last major German oil company, Deutsche Erdol AG, passed into American hands by being taken over by Texaco.

Despite possible justification of the government interest in Veba, for reasons of security of supply, better access to industry information or for better government to government dealing with producer countries, the stated government policy in oil matters was still to rely on market forces. Germany was the only European country without some kind of price control on oil products. Certainly the merger of the two companies promised economies of scale, rationalization of facilities, and the creation of a German national group of comparable size to the foreign multinational companies operating in Germany; but at the beginning of 1980 the future role of Veba in the domestic oil market was paradoxical. In view of the varied interests of the merged company there was little question that the new group would be commercially viable, but as a national oil company Veba distinguished itself from its European counterparts in having very little crude oil of its own and no clear political framework in which to act. Indeed the government, in creating an exploration company in its control, appeared in the process to have obtained an integrated national oil company without fully working out the implications of sustaining this company in a freely competitive market for petroleum products. This paradox was made even more pointed by the partial merger of Veba with the German affiliate of British Petroleum in 1979. While the perception of Veba as the German national oil company continued it was somewhat less certain what its actual role really was.

Veba AG and Gelsenberg AG – Their Early History

Veba AG was incorporated in 1929 when the Prussian state transferred to it the interests of the state in coal mining and electricity supply while retaining ownership of the company. Among the assets acquired was a mining company, Bergwerkgesellschaft Hibernia, founded in 1873 and which had become one of the biggest mining companies in the Ruhr, and Preussiche Elektrizitats AG, which had been established in 1927 to promote electrification of large areas of Central and Eastern Germany.[1]

By 1965 in addition to its coal and electricity interests Veba had moved into petrochemical production, glass making, trading, transportation, and other services as well as having interests in nuclear energy. At this time, in keeping with the government's policy of spreading ownership of public corporations, Veba sold off 56.3 per cent of its shares to 1.2 million shareholders, thereby giving it more shareholders than any other European corporation. In 1968 in the face of readjustment problems in the industry the government concentrated all Germany's coal mining interests into a few large regional companies. Veba AG was the largest shareholder in the principal company, Ruhrkohle AG, with a 14 per cent share. A year later, in 1969, in an equally significant move Veba AG, together with seven other independent Germany companies, formed an oil exploration and production company, Deminex, in which Veba's share amounted to 18.5 per cent. By the beginning of the seventies Veba AG was a large company with extensive energy interests together with significant activities in a number of fields, not all directly related to energy.

Gelsenberg AG, although indirectly dating from the 1920s, was a younger and smaller company than Veba, having been established in 1953 as one of the successors of the 'Vereinigte Stahlwerke AG' group.[2] Gelsenberg had diversified interests which encompassed electricity production and distribution, exploration, and production of hydrocarbons and nuclear materials, organic- and petrochemicals, general trade in iron and steel products and in wood and building materials, and the mining and processing of coal. Gelsenberg also had a 25 per cent interest in a gas-producing and distributing company, Ruhrgas AG, which supplied about a quarter of all German natural gas consumption. As with Veba AG, in 1968 the coal mining interests passed into the hands of Ruhrkohle AG with Gelsenberg AG retaining a 13.1 per cent interest, and a year later Gelsenberg was another investor in Deminex, also holding an 18.5 per cent share. Thus, although smaller (1970 turnover DM3.5 billion; Veba AG, DM8.1 billion),[3] Gelsenberg AG was a company similar to Veba AG, although such differences as existed were significant (see Appendix 1).

The electricity activities of Veba were ten times the size of those of Gelsenberg, and Veba's petrochemical operations were twice the size; both had highly developed transportation and trading companies (Gelsenberg – Raab Karcher GmbH, Essen; Veba – Hugo Stinnes AG, Mulheim/Ruhr). Veba had lignite mining and hollow glass activities, Gelsenberg had natural gas production facilities in Holland, and both companies sold petrol and diesel through a distribution network in Germany under the trade name ARAL, in which their respective shares were 28 per cent. Gelsenberg had recognized experience in oil matters even to the extent of being asked by the Libyans in March 1971 to negotiate on behalf of all the oil companies operating in Libya over nationalization.[4] With their interests in Deminex, Ruhrkohle, ARAL, and various other joint ventures, the companies were not strangers to one another and in 1969 a merger between the two was suggested by Veba, but was declined by Gelsenberg's chairman, Walter Cipa. The subsequent interests of the government in bringing the two together was then unforeseen.

Economic Policy and Energy Policy

In the fifties and sixties Germany followed a policy of the 'social market economy'. The basic elements of this policy were private ownership of the means of production, freedom of entrepreneurial initiative, unrestricted competition, and the guarantee of certain social stability.[5]

Much of the sentiment of these elements were incorporated in the Act Against Restriction of Competition, which was generally called the Cartel Act (1957). However, the government favored oligopolistic mergers if, by doing so, it felt that the chances for stability and growth were improved. A result of this was the foundation in 1968 of Ruhrkohle AG, in which both Veba and Gelsenberg had participated. Implicit in the act was a provision that governmental subsidies should be cut off from those companies 'which do not prove to have firm-size desirable for the achievement of optimum productivity.'[6]

In 1972, Karl Schiller, the then Minister of Economic Affairs, introduced an Act for the Promotion of Stability and Economic Growth, and so the social market economy progressed to the 'enlightened market economy'. The purpose of this 'enlightened market economy' was 'to develop a compatible system which could reconcile systematic consideration of national problems with economic freedom and with an international outlook on the part of the socio-economic groups and economic units'.[7]

This then set the pattern in the energy industry for relations with the Economics Minister and the Cartel Office. The government appeared to be sponsoring concentration, but in fact it was precisely this which had become its primary concern before cartelization. The government was on the horns of the dilemma which traditionally emerge when evaluating the level of concentration necessary in any one sector of industry: too much concentration put excessive monopoly power in the hands of the sellers and the buyers suffered – insufficient concentration failed to provide the security and stability necessary, both in high technology and in declining industries, to invest and adapt to the best advantage of the consumer. Not only was concentration becoming a problem in Germany in the late sixties, but also there was sectoral instability, where the interests of the sectors were represented by powerful lobbies and the government was placed in the middle. Considering the fact that producers' associations representing sectoral interests were relatively efficient units for generating information with respect to their own sectors, they had more force as a rule than an individual firm. Their collective financial resources and their ability to exercise collective bargaining power in the interests of the sector were also superior. In oligopolistically structured industries, the limited number of firms and their habit of cooperation added to their strength. Where these industries were in decline, the government found itself being pressured into supporting them, hoping that by doing so it would generate political strength rather than political loss.

Strategy formulation in the field of energy policy faced constraints other than anomalies at the national economic level posed above. Oil produced in Germany had supplied one third, i.e. 3.3 million tons, of domestic consumption in 1955 but although local production doubled by the early seventies, consumption had increased 15 times.[8] Coal, Germany's principal energy source, was facing competitive pressure from oil. Because of the industrial and social problems raised by this competition the government had instituted a coal protection policy, the origins of which went back to the late fifties. (See Appendix 2 for German energy consumption.) The policy was designed to attenuate the advance of oil where it cut most directly into coal's market – first heavy oil, followed later by light or home-heating oil – through a heavy excise tax on these oils.[9] (The tax was imposed for fiscal reasons as well.) Also the federal government had introduced a protective tariff on crude oil imports, which was meant to protect inland crude production and which was abandoned only reluctantly in 1964 as a concession to Common Market partners in order to shape the common external tariff system. There had been a tax preference for crude oil processing (lasting until 1964) in the former coal–hydro–generation plants where crude was refined into petrol exclusively. There was a special tax on heating oil, introduced in 1960 which was still in effect in 1977. There had also been a short-lived (1958–59) effort to regulate the oil 'invasion' by having the principal domestic and foreign firms in coal and oil form a coal/oil cartel, followed by a temporary successful effort to encourage the oil companies into voluntary acceptance of a scheme limiting the growth of heavy and light fuel oil supplies to certain annual percentage rates. Finally laws were passed (in 1965, 1966, and again in 1974 and 1975) to subsidize conversion to coal (see Appendix 2 for energy balance and details of oil consumption).

Consequently, heavy fuel oils were at a competitive disadvantage *vis-à-vis* coal, and the less a company produced, the better off it was. The government was anxious, however, to ensure that security of the existing small domestic crude base was amplified by having a crude base of some size under the management of German companies. To this end it encouraged the formation of Deminex in 1969 by eight independent German companies including Veba AG and Gelsenberg AG (see Appendix 3). Only Veba AG and Gelsenberg AG had oil refining and marketing operations. The objective of Deminex was to supply crude oil to an independent German oil industry and the initial objective was to supply 25 per cent of its shareholders' requirements by the middle of the seventies.[10]

The Merger

As the seventies started the economic and energy policies of the German government had a number of conflicting pressures exerted upon them. The Christian Democrats, who had tended to ally themselves with big business, with its tendency to form cartels, were replaced in 1969 by a coalition of Social Democrats and Free Democrats who, paradoxically, championed free enterprise and were more active in anti-merger activity. Nonetheless in the field of oil policy the new government proceeded along the line indicated before, which was leading toward the creation of a national oil champion.

The creation of Veba–Gelsenberg coincided with the period of increasing instability in world oil markets culminating in the 1973 energy crisis when the German government felt itself particularly vulnerable. The initiative to create a national champion through a concentration of strengths and resources was taken beforehand. Following a meeting in May 1973 between the head of Veba AG and the Economics Minister, Hans Friderichs (who had only just replaced Helmut Schmidt), it was announced on June 18 that Veba was to take over Gelsenberg.

The Chief Executive of Gelsenberg, Walter Cipa, appeared to be quite unaware of the government's intentions, having been occupied with a scheme which, though not perhaps creating a national champion, would build on his long-established contacts in the international energy business to create an important German presence in the oil industry. Two days after the announcement, Cipa went to Friderichs to offer his alternative plan which, in his opinion, would cost the Germans less. Cipa's mission did not meet with success. On November 28 the Bundeskartellamt was notified of the merger, which in January 1974, it declined to approve. The government, however, had by this time decided that the merger should go through, and with governmental involvement.

Prior to December 1973, 48.3 per cent of the share capital of Gelsenberg had been held by Rheinisch–Westfalisches Elektrizitatswerk AG. The government acquired these shares and by January 23, 1974, held some 51.3 per cent of Gelsenberg's share capital. The purchase was undoubtedly made easier by the fact that Gelsenberg had been only marginally profitable every year since 1970 (and was to continue to be so until completely merged with Veba). On February 1 Friderichs announced that he had decided to approve the merger under the general interest paragraph of the law against restrictions to competition (GWB). The Bundeskartellamt's objections were replaced by a decree of Friderichs', and on May 13, the shares acquired by the government were given to Veba AG. Restructuring and consolidating of the shareholding continued through 1974 and early 1975. Finally, in 1975, Veba AG acquired 96.1 per cent of the equity capital of Gelsenberg AG and concentrated the activities of both companies in the petroleum and chemical fields into Veba-Chemie AG, a 100 per cent owned subsidiary operating company. The participation of the German government in the merged company became about 44 per cent, and it started operating as a single entity on January 1, 1975,[11] with Veba serving as a holding company for Veba-Chemie AG and the other Gelsenberg and Veba companies operating in different fields.

Not only had the government assisted politically in merging the two companies, it had also done so financially. The gross amount involved necessary to amass the government's majority shareholding was DM800 million. However, the net amount was only DM140 million, the sum over the market value of the shares which the government had to offer to Gelsenberg shareholders to exchange their shares for Veba stock.

Deminex and Oil Supply

Perhaps most important of the changes in the shareholdings that took place as a result of the merger was the controlling interest acquired by Veba in Deminex. The interests of two of the companies owning Deminex, Preussag AG and Deutsche Schachtbau, were acquired by Veba in September 1974, and Gelsenberg's was consolidated with that of Veba-Chemie. The total Veba holding consequently amounted to 54 per cent, and although Deminex operated as a private company under contract to Veba it could now be regarded as Veba's exploration division.

Deminex would continue to receive considerable government subsidy for overseas exploration. Between 1969 and 1974 it received DM575 million and a further DM800 million was envisaged between 1975 and 1978.[12] It seemed likely that the subsidizing, which consisted of a loan for 75 per cent of the total amount paid for exploration (which did not have to be repaid in cases where no oil was found), would continue after 1978. If oil was discovered, DM3 per ton (a very inexpensive rate) was to be paid back to the government. If Deminex bought into an existing proven field, the government provided 30 per cent of the purchase price. It was not intended that the actual capital required to finance the development of potential hydrocarbon reservoirs would come from the budget. Like nearly all other major and minor oil companies, Deminex would have to go to the money markets to raise the capital for such projects.

In its first years, Deminex had concentrated most of its effort on acquiring participation rights in existing exploration or development situations. Later it had preferred to obtain exploration rights in new or underexplored areas, especially offshore[13] (see Appendix 3 for budget allocation). In the Spring of 1975 Deminex turned to the North Sea and acquired the participation rights of United Canso Oil and Gas (UK) and then those of Champlin Petroleum in a 250 km^2 block which contained by far the largest portion of the Thistle field (discovered first in 1973). Production was expected by the end of 1977 from recoverable reserves estimated to be 60–70 million tons.[14] The total share of Deminex in the block amounted to 42.5 per cent. While concentrating as much of its activity as the oil reserves warranted near to Germany, Deminex was very active in other parts of the world, both alone and in partnership with most of the oil majors and national oil companies. Some difficulties had been encountered, notably BP's attempt at a joint partnership with Deminex in Abu Dhabi which fell through because of governmental indecision. Also Deminex had had limited success with NIOC in Iran.

It was perhaps too early to judge whether or not Deminex had been successful except that its efforts were, so far, not cost-effective. By the end of 1979, the oil requirements of its shareholders were only covered for a small part by its own resources although Deminex hoped to improve this position substantially on a long-term basis. It might have been possible for Veba to acquire its crude supplies more cheaply from the majors, but then the 'national interest' would not have been satisfied. Also, since the majors were uncertain as to the security of their own supplies, they were probably reluctant to commit themselves to any long-term arrangement with Veba or Deminex. Veba had used other means of securing crude supplies, namely in direct government deals, as had been the case in Saudi Arabia. In late 1973, the Oil Minister, Yamani, had told a German magazine, 'You must industrialize Saudi Arabia. Then we shall give you as much oil as you need.'[15] Twelve million tons of Saudi participation oil were subsequently allotted to Veba over the next three years, to be delivered for the most part in 1976–77 as the government tried to take advantage of the decision of the Saudi government to deal with national oil companies such as Veba, in addition to MNCs. Other supplies of crude came from Algeria, the USSR, and Libya (see Appendix 4).

Refining

The crude oil supplied to Veba went to three fully owned refineries, three with approximately half ownership, and one where Veba's share was 25 per cent. Total capacity in 1976 was 25.9 million tons (see Appendix 5) but in 1975 it had been seriously underutilized at 60 per cent of capacity[16] (4 million tons capacity lay idle). The figure was typical of the industry as a whole, which had seen capacity utilization drop to 62.2 per cent in 1975 from 83.4 per cent in 1973.[17] Refinery investment was as uncontrolled as marketing in Germany's free market conditions and overcapacity was a constant preoccupation even in the boom years before the 1973 crisis. With expected demand for oil in West Germany conservatively estimated to grow at a rate of less than 2 per cent per annum until 1985,[18] for which there was adequate refining capacity (inclusive of those projects underway), 1973 levels were not expected to be attained again until 1980. In light of this situation some companies had reduced their capacity. For example, Mobil closed its Bremen refinery in 1974, Fina closed at Mulheim, Elf at Essen, and Veba at Gelsenkirchen. At the same time, a Mobil refinery had gone on stream at Wilhelmshaven and Elf was expanding its refinery at Speyer.

Investment in new refinery capacity could only be justified if it altered the mix in output to produce lighter products of which Germany was in short supply. Veba possessed 19.7 per cent of total available German capacity (153.9 million tons) in 1976, and Veba's refineries had a total of 3.6 million tons of cracking facility in 1976, most of which had come from Gelsenberg and which represented

19 per cent of the total cracking facilities in Germany available in 1976. A further 1.1 million tons of conversion capacity was being constructed (1.6 million tons when including a fuel oil gasification plant to produce ammonia and urea).

The significance of this was that without cracking facilities it was difficult to avoid producing excessive quantities of the heavier derivatives with which the market was already too well supplied, especially since the power stations were compelled to use coal under the legislation preventing the conversion from coal to oil. Even BP, which had recently increased its market share in Germany, was heavily represented with fuel oil and had lost approximately $150 million in 1975. Veba too was heavily represented in fuel oil, but, unlike BP which was directly exposed to the extreme competition of the fuel oil market, Veba with its over-supply at least could envisage using its fuel oil in its power stations if the coal protection policy were ever to be lifted.

The bulk of the distribution of refined products was undertaken by ARAL, although Veba also possessed a small, fully owned network, FANAL, purchased from Gulf in 1973. The principal shareholders in ARAL before the merger were Veba, Gelsenberg, and Mobil, each holding 27.96 per cent of the stock.[19] The merger gave Veba a controlling interest (almost 56 per cent), but the by-laws of ARAL stipulated agreement with Mobil, which had no direct German outlet, on all major issues. Mobil had been associated with ARAL since before World War II and by arrangement had supplied 3.5 million tons per year to Veba and Gelsenberg's refineries.

The arrangement was renewed until 1980 following the merger as a 'short-fall' contract for up to 3.5 million tons at preferential prices. Mobil had previously been the main crude supplier of Gelsenberg, who had offered Mobil a share in ARAL. By agreement ARAL was the outlet for the petrol and automotive diesel products of Veba, Gelsenberg, Mobil, and Wintershall, the other shareholder of ARAL with 15 per cent of the equity. ARAL was the market leader for petrol, holding 25 per cent of the market in 1975, with its nearest rival, Shell, holding 16 per cent. But ARAL achieved its sales less efficiently than the MNCs (see Appendix 6).

The Post-1973 Environment

The immediate aftermath of the 1973 energy crisis was a very steep rise in crude oil prices in Germany, moving from DM85.20 to 223.86 per ton between 1973 and 1974.[20] In line with Germany's free market philosophy, prices were set by supply and demand and, when the OPEC crisis was at its height, oil prices in Germany were the highest in Europe. The government was convinced that Germany was supplied because the prices went up dramatically but this was questionable because, in fact, the whole of Europe was supplied, including France where the prices were strictly controlled.

One effect of the worldwide rise in crude oil prices was that those German producers having domestic natural gas and oil production (all of them oil MNCs) began to make extraordinary upstream gains (see Appendix 7). This enabled them to subsidize losses which occurred in 1975–76 in downstream operations, and by 1976 the difference between the companies having producing operations in Germany and those which did not was beginning to distort the market. Industry sources estimated that the cost of production of both gas and oil was about $4–5 per barrel of crude equivalent including existing royalties and taxes, but the market price was about $12–13 per barrel.

To equalize the competitive situation or to create a 'fairer' market condition, some members of Parliament argued in early 1976 that a windfall profits tax should be introduced on German domestic production. Veba and BP, neither of which had domestic production, would have been the indirect beneficiaries of such an action. This was obviously resisted by the domestic producers and Minister Friderichs; political factors made it seem unlikely that a windfall profits tax would be introduced and the political decision in turn was influenced by legal considerations. Furthermore, Friderichs and the Social Democratic and Free Democratic Parties seemed ambivalent on the subject. The Christian Democratic Party preferred to champion the coal industry, as it had always done in the past, and of course, higher oil prices would benefit the coal industry. But the chemical industry foresaw being confronted by a powerful Veba (assuming it was given – directly or indirectly – the windfall tax proceeds) which might upset the longstanding supply agreements that the chemical firms had built up with the oil MNCs. Further contrary arguments against the windfall tax suggested that it was against the system of market economy and did not make the supply of energy any safer. Protagonists of the windfall tax envisaged benefits of revenue for the government through taxation, and revenue for the state of Lower Saxony, which had the greatest share in oil and gas deposits, although the proposed tax was on the federal level and Lower Saxony would have had to obtain its share through the political process.

The situation had added relevance to Veba due to the fuel oil problem. The sharp rise in prices and the projected diminution of oil in the energy demand from 52 per cent of the total in 1975 to 42 per cent in 1990[21] had already served to turn down fuel oil consumption. Like BP, Veba suffered serious losses in its petroleum activities in 1975[22] and lesser, but still substantial, losses in 1976, which nullified to a large extent the good profits made in other activities. Veba, operating only in Germany, was vulnerable to the changes in the market due to its lack of domestic production and reliance on fuel oil and it did not have the ability to balance its production to satisfy the needs of several markets. Its transportation costs were high as it did not have the economies of scale MNCs could derive from global marketing and production activity.

Of the oil MNCs, Esso and Shell appeared to be particularly well placed. They had natural gas and oil production in Germany, and they had refinery facilities in Holland from which oil could be imported into Germany while the 'heavy' products could be sold outside Germany. They both had significant market shares and were particularly strong in petrol. They had established relationships with the German chemical industry and could benefit from their highly effective central planning and other service facilities geared to maximize worldwide efficiency. In

1975, trade sources had estimated that the German oil industry had lost DM2 billion at DM20 per ton, much of it by BP and Veba. Gulf and some American independents had already pulled out, and ENI was possibly in the process of withdrawing. The reaction of the companies to the free market conditions in Germany was varied. Esso was strongly against any action to curtail the 'openness' of the market, Shell had come out for a kind of modified 'open market', and BP and Veba were urging the government to adopt some kind of controls. The variety of reaction was symptomatic not only of the variety of condition experienced by each competitor in the oil market but also of the contradictions of the government both in its energy policy and in its attitude to the market mechanism.

Veba – Assessment and Outlook

The newly formed Veba, not unsurprisingly, found itself with no clear identity. When developing its policy in the early seventies, and in assisting and promoting the merger, the government had been motivated by a number of objectives. One was certainly the promotion of a reliable source of crude oil under the control of German companies. Through the merger, Veba acquired a majority interest in Deminex thus, hopefully, strengthening Germany's efforts to find more crude oil under German control. Moreover, by combining Veba and Gelsenberg, a group of sufficient stature would be created which could deal with both the oil majors and the producing countries. By the government having 44 per cent equity the standing of the company and that of the Germany market itself would be enhanced. As the government became more involved in the market, it would acquire a window into the industry which would allow it to formulate policy and exercise control. Another motive in creating a stronger Veba almost certainly was to arrest the fast declining German presence in the German oil products market (see Appendix 8). At the purely industrial level, the government hoped the merger would create economies of scale (which would increase competition) and provide integrated management and better use of capital, all of which might imply that the government also looked to receive a reasonable return on its not inconsiderable investment, particularly in the case of Deminex. In the light of these objectives Veba's and Deminex's achievements by the end of the seventies were limited.

The combined interests of Veba and Gelsenberg dated back a long way, and in the late sixties had benefited from a government in favor of oligopolistic mergers. Most important of the joint ventures was their cooperation in Deminex, the exploration venture. At the time of the merger it had been felt by the government that the ownership by eight companies was not conducive to achieve Deminex's goals because not one of them had a controlling interest in Deminex and also because at least four of the companies lacked both downstream operations (and hence the need for oil) and also, in some cases, the financial resources to explore for oil. The merger and subsequent rationalization of the shareholding gave Veba its (56 per cent) controlling interest. The exploration aspect of the oil industry must have seemed increasingly important in the early seventies and particularly urgent in 1973 and 1974, and the strengthening of Deminex's efforts and, therefore, Germany's, was probably the primary reason for the merger.

Despite the importance of exploration in an integrated industry other phases of operations were equally important, of which the most immediately related was supply. With the help of the government, Veba had negotiated directly, and successfully, with some of the oil-producing countries. Accordingly it had achieved some success in securing management of its crude requirements; the rest had to be bought at commercial prices. Since the oil MNCs still realized a 22–25 c/bbl crude profit on their OPEC production, this was the amount that Veba did not have on virtually all of its crude. Veba was a very long way from being adequately supplied by German-owned and -controlled crude sources. Thus while the upstream activities of Veba had been the original justification for the merger they were too new and undeveloped to have brought noticeable benefits.

In retrospect, following the 1973 crisis it seemed that the downstream operations of the industry had received insufficient consideration at the time of the merger. Refining seemed to pose the greatest problems in the short and long term if Veba could not somehow generate enough money to convert its refineries for lighter products. The scale of the changes in the energy market and their effect on refinery capacity was unforeseen by the German government, as by everyone else, and was no doubt a contributory factor to why the government appeared to have given insufficient thought to the downstream implications. If the problems of running profitable refining operations were common to most, if not all, operators in Germany, the advantages of having German oil and gas production were open only to some. Veba, with no domestic oil or gas production, made no upstream profits, and in this sense Veba had argued that its government did not give it equal treatment with the MNCs. In France and Italy, the oil industry had been regulated in such a way that the national companies obtained the upstream profits from protected domestic oil and gas production. In Germany, in contrast, the national company received no protection. Despite the leverage the windfall profits gave to competitors, the government had refused so far to base its energy program on the company and much of 'countervailing power' had become redundant. Friderichs, who was after all the architect of the merger, was also the principal voice in the fight to protect the free market and consequently the government did not give Veba preferential treatment. However, Veba needed to look to the government for assistance since its environment was not the best for survival, let alone prosperity.

The saving grace of Veba was that it was a conglomerate and there was therefore the potential for cross-subsidizing the losing oil operations. From 1975 to 1977, Veba's electricity activities had subsidized its oil and gas activities, but there was probably a limit to which Veba's electricity division was able and willing to cross-subsidize the unprofitable oil operations. Veba had decentralized management and such a cross-subsidization for a long-term period would run against managerial discretions. Moreover, electricity rates were publicly controlled and it was unlikely that the individual states would agree to higher electricity rates in order to subsidize oil operations for any appreciable length of time (see Appendixes 9 and 10 for financial results).

Economies of scale that might result from the merger by mid-1977, only two years after the merger, had not appeared. The activities grouped under Veba-Chemie AG, principally oil and chemicals, were strongly centrally managed, as were similar corporations in Germany, but the benefits from restructuring of so large a group would take time to appear, and outside Veba-Chemie AG central management was less strong. Transportation needed integrating, the trading companies were still separate, and accounting practices were not yet completely aligned. (See Appendix 11 for company structure.)

If Veba's losses continued or increased, then the government would probably be forced to take action. Though holding some 44 per cent of the shares, there were 1.2 million shareholders who could embarrass the government if it failed to protect them. One possible measure would be for the government to forego its dividends while the private shareholders would continue to receive them. Alternatively Veba could be exempted from the requirements to burn coal in its power stations and could thereby consume its fuel oil surplus, which would be a device through which the entire Veba system could be better integrated.

One achievement of the merger was that the government had a window into the industry, although just how much they were looking through it was unclear since they had been following very much of a 'hands off' policy since the merger. If the government was able to see its way to thinking postively about Veba and its commitment to a 25 per cent German-controlled stake in Veba's crude supplies, then at the very least, the limited governmental assistance programs to Deminex would have to continue. Government assistance would also continue to another important Veba investment, Ruhrkohle, which had received an average subsidy on production of DM23 per ton in 1975, which was an indirect aid to Ruhrkohle's shareholders.

The key to Veba's future and the government's attitude toward Veba would seem to depend on the business cycle. With economic activity improving, so would the demand for oil. The composition of the product mix would also improve; stronger demand for naphtha and petrol would bring corresponding price increases. In such a cyclical situation, Veba's electricity operations would improve considerably and so would, though to a lesser extent, its oil activities. However, if the boom were to run its course and be replaced by another significant decline in economic activity, Veba could find itself in serious difficulties and in need of further governmental assistance. In such a case, it would be reasonable to predict a change in government philosophy which would allow it to change the market environment in a similar way to that employed in 1968 with the coal industry.

The Interim Lessons

The lack of political guidance for Veba by the government was of major importance in the assessment of Veba's past and speculation about its future. After partially denationalizing Veba in 1965 by selling 56 per cent of its shares to the public, the government, eight years later, had come upon the view that it wished to have a national champion, and an enlarged one at that, after all. But having assisted in Veba's merger with Gelsenberg, it appeared that the government had wanted to step back and leave Veba's management to sort out its fortunes. No further sizeable government funds had proved forthcoming, except for Deminex, and Veba was expected to survive in the market. Being national champion had not brought any preferential treatment and the government had not provided Veba with a competitive advantage over other oil companies operating in Germany. In that sense, Veba was not comparable to such national oil companies as Elf-Aquitaine and ENI. However, the government still held 44 per cent and could be expected to invest in Veba if only to cover its downstream losses which every downstream operator in Germany was incurring; in this sense, it would be behaving like a private owner. At the same time, after the merger the government had produced an adjusted energy program (October 1974) which was in no way based on Veba. In March 1976, when Dr. Engelmann of the Economics Ministry was asked if Veba would be used as an instrument of government policy, he reaffirmed that the government energy policy would not be based on Veba. The government had felt that Veba and Gelsenberg should be merged, had brought about the union, and then found its policy to remain uninvolved brought into question.

Given the very competitive situation in the German market, because of its openness and proximity to Rotterdam and the very large amount of imports of oil products (see Appendix 12), Veba would, it seemed, need government support to survive, let alone prosper. The key, therefore, was government policy and two years after the merger no such policy seemed to be emerging. One reason for this was that Veba, until the present at least, did not have a powerful lobby in Bonn. The second was that the government had been caught short by its own free market ideology and now was unable to execute a turn-around. This was not to say that governments needed to create national oil companies. It was unlikely that the British, who have also taken an open market stance, would have created their national oil company had not oil been found in the North Sea. The Japanese were probably served as well by strong control over the market operators by the government, without a national oil company. Nevertheless, the German government, probably following the French and Italian examples, having decided to create a national oil company, had not created the necessary political and economic framework that were the logical consequences of such a decision. The French and Italian governments had been fully aware for the last few decades that they wanted a national oil company and that such enterprises, being late-comers in a mature oligopolistic market, needed sizeable and consistent assistance which they wanted a national oil company and that such enterprises, being latecomers had been consistent since 1928. In Italy, ENI had been powerful enough to steer government policy in its interest since the early fifties. The awareness of the need for support had, it seemed, not yet struck the German government and whether it would or would not was a matter of conjecture. In Germany, Veba had not been strong enough to influence government policy and the government itself seemed to be at least of two minds as to what to do with Veba, with the result that so far it had done nothing. In the final analysis, the crucial question was whether Veba could be made viable without the government's 'spiritual' and policy backing. It seemed doubtful.

While German energy observers were nearly unanimous in their view that the French and Italian national oil company experiences 'can never happen' in Germany, in the creation of the merged Veba the government had created a sizeable potential vehicle for intervention in oil matters if it so chose.

In the final analysis, the most important internal factor working on Veba would be the fortunes of Deminex, which had been the key to the merger. Deminex's record in supplying Veba with secure and inexpensive oil could only be characterized as poor and unless there was a drastic improvement in Deminex's performance, this fact, together with a lack of government identification with Veba, could only bode ill for the company.

The question of Veba for the German government actually went beyond the potential fortunes of that particular company. It brought into focus the question of whether or not it was possible for Germany to continue to have an orderly market in oil when virtually all other countries in Europe had some kind of control over their fuel markets. With Germany being the only large 'free' market in Europe, coupled with the proximity of Rotterdam with its export refineries which were mainly poised to supply the German market, the fate of a purely German company, such as Veba, was made all the more untenable. The future of Veba, then, was a classic illustration of the conflicts that may arise between a government's 'market' philosophy and state intervention because of different perceptions of public and private interests. The case was particularly poignant because the conflict was essentially unwanted and unforeseen by the government. The government had wanted to step up German oil exploration efforts, a relatively simple exercise in a very complex business. Instead it had acquired a very large integrated oil company with all of its complexities, which was situated in the world's most competitive market.

The Second Merger and Some Further Lessons

Veba continued to register losses in its oil operations in both 1977 and 1978. The other major company in the same financial predicament was the German affiliate of BP, mostly because the latter increased its market share – and mainly in fuel oil – at the time when oil consumption slowed down. Veba's losses were, of course, also due to the slow-down in the economy making the excess capacity problem both in refineries and tankers more serious. Moreover, what was worse for Veba, the profit on its electricity operations decreased over 1976 while the losses in the chemical division increased substantially. While 1978 showed better results than 1977, the petroleum division was still operating at a loss. (See the 1977 and 1978 Annual Reports.) In the face of these losses, Veba looked to Bonn and Bonn continued to look the other way. This was not the only governmental option; it could have, for instance, followed the example of the Christian Democrats in 1969 when the government 'rationalized' (aided) the coal industry. But in a political paradox, the Social Democratic coalition of 1978–79 preferred a free enterprise solution. Veba thus was left to work out its own economic salvation and it managed to do so despite some governmental objections.

One obvious free enterprise solution for Veba was to join forces with an oil multinational and preferably one with adequate crude supplies of its own. As far back as 1975, Veba negotiated with Gulf for just such an arrangement.[23] While the crude arrangement never materialized, Veba did purchase Gulf's petrol station network. In the summer of 1978, however, Veba's efforts met with success as it established a series of joint ventures with BP (Germany). Effective January 1979, BP (Germany) acquired the original Gelsenberg operation, Veba's FANAL petrol station operations (the ones it acquired from Gulf), substantial portions of Veba's diesel, fuel oil, and coal marketing outlets, and Veba's share of the Speyer and Ingolstadt refineries. The agreement also included BP's acquisition of Veba's 25 per cent share in Ruhrgas – a matter which became a subject of governmental concern. In exchange Veba received about DM800 million and a twenty-year supply contract of 3 million tons of crude per year.[24] While 3 million tons per year does not seem to be much, it was enough to the crude short Veba and also there was an implicit understanding that BP would assume the role of the friendly supplier in case Veba experienced a shortfall. Overall, both companies realized, through the merger, economies of scale in refining, transport, and marketing. Finally, Veba announced that the bulk of the DM800 million it received would be committed, through Denimex, to solve Veba's chronic shortfall of its own crude.

The Cartel Office blocked the merger,[25] just as it did the original Veba–Gelsenberg merger in 1974. (The objection was then overridden by the Economics Minister.) History repeated itself and in March 1979 the Economics Minister overruled the Cartel Office's objections and approved the merger.[26] In order to obtain approval BP surrendered a part of its newly acquired voting rights in Ruhrgas.

According to the 1979 Annual Report, Veba recorded a profit increase of some 150 per cent over the previous year. This increase was obviously not yet due to the merger but rather represented – in addition to an increase in sales because of the improved economic conditions – inventory profits realized as the result of the large OPEC price increases in the course of 1979.

Financially, the merger was favorable to both Veba and BP. Veba bought Gelsenberg for DM800 million and received the same amount for a portion of it six years later. It is estimated that the portion remaining with Veba was worth about DM900 million.[27] These portions include the Gelsenberg shares in ARAL and in Deminex. Last but not least, Veba managed to rid itself of its excess refining capacity.

The dilemma facing the government was resolved by the government giving up, as it were, having its national oil company. After six years of frustration and indecision, it is not clear whether the government regarded the merger with BP with relief or chagrin. It is also not clear whether when Veba's Chairman said that 'our aim was to create a German oil group on the basis of a free market without *dirigiste* tools and (that) we succeeded'[28] he was expressing a political and business judgment or the acknowledgment of a *fait accompli*.

Appendix 1. Veba AG and Gelsenberg AG – 1973 Results[29]

	Veba AG	Gelsenberg AG
Crude oil production (thousands of tons)	–	2,054
Crude oil throughput (thousands of tons)	10,832	7,597
Natural gas production (millions of m^3)	–	164
Petrochemical products (thousands of tons)	850	478
Electricity generated (millions of kWh)	41,379	4,267
Lignite production (thousands of tons)	9,377	–
Petrol production (thousands of tons)	3,044	2,034
Middle distillate production (thousands of tons)	4,001	2,295
Fuel oil (thousands of tons)	3,092	1,715
Hollow-glass production (thousands of tons)	576	–
Sales (DM million)	12,467	5,548
Profits after tax (DM million)	402	49
Investments (DM million)	1,680	222
Depreciation (DM million)	1,030	270
Employees	58,776	14,579

Note: Coal production is not included as Ruhrkohle AG (Veba 14.1%, Gelsenberg 13.1%) is not consolidated.

Gas sales by Ruhrgas AG (Veba 5.2%, Gelsenberg 25.0%) are not included for the same reason.

Appendix 2

Germany – energy consumption and supply[30] (million tons of coal equivalent)

		1970	1975	1980	1990
Total		336.8	345.2	410.0	540.0
of which (per cent)	Coal	38	30	26	20
	Oil	53	52	49	42
	Natural gas	5	14	17[a]	17
	Nuclear power	1	2	6	19
	Hydroelectricity	3	2	2	2

[a]Indigenous production in Germany and imports from the Netherlands, USSR, Norway, and Algeria.

Germany – oil consumption (million tons of oil)

		1970	1975	1980	1990
Total		124	125	141	158
of which (per cent)	Petrochemical feedstocks	9	8	11	14
	Petrol	12	15	15	15
	Diesel, automotive	8	8	8	8
	Light heating oil	35	36	35	32
	Heavy fuel oil	21	18	16	15

Appendix 3

Original and final shareholdings in Deminex[31]

	1969 (per cent)	1976 (per cent)
Gelsenberg AG	18.5	54.0 (Gelsenberg AG and Veba-Chemie AG combined)
Veba-Chemie AG	18.5	
Wintershall AG (100% owned by BASF)	18.5	18.5
Union Rheinische Braunkohlen Kraftstoff AG (25% owned by Hoechst, 75% by RWE, a utility)	13.5	18.5
Deutsche Schachtbau GmbH	10.0	–
Saarbergwerke AG (coal and electricity)	9.0	9.0
Preussag AG	7.0	–
Deilmann AG	5.0	–

Distribution of exploration budget of Deminex, 1969–75

The budget of DM1,835 million has been allocated as follows (per cent):

Middle East	18	
North Africa	14	
West Africa	10	
Latin America	10	
North Sea/North Atlantic	20	
Others	4	
Projects under negotiation	10	86
Unallocated		14
		100

Appendix 4. Veba's and Germany's Crude Oil Supply

Veba's crude oil sources, 1975[32]

	Millions of tons	Per cent
Algeria	4.4	27.5
USSR	2.8	17.5
Saudi Arabia	2.2	13.8
Libya	2.3	14.4
Nigeria	1.7	10.6
Iran	1.3	8.1
Other countries	1.3	8.1
Total throughput	16.0	100.0

German crude oil imports by sources, 1975[33]

	Millions of tons	Per cent	
Middle East	46.8	52.0	
of which, Saudi Arabia	18.6		20.6
Iran	14.2		15.8
Venezuela	2.2	2.4	
North and West Africa	37.4	41.5	
USSR	3.1	3.4	
Norway	0.6	0.7	
	90.1	100.0	

Appendix 5. Veba and German Refinery Capacities[34]

Veba, 1976

	Total capacity (millions of tons)	Veba share (per cent)	Veba capacity share (millions of tons)
Gelsenkirchen–Scholven	10.0	100	10.0
Gelsenkirchen–Horst	7.0	100	7.0
Oberrheinische Mineralolwerke GmbH, Karlsruhe	7.0	45[a]	3.2
Erdolraffinerie Ingolstadt AG, Ingolstadt	7.0	50[b]	3.5
Erdol–Raffinerie Neustadt GmbH & Co., oHG, Neustadt/Donau	7.0	50[c]	3.5
Erdolwerke Frisia AG, Emden	2.4	100	2.4
			29.6
Elf–Gelsenberg oHG, Speyer	7.0	25	1.7

[a]Texaco held 45% and Conoco 10%. Modernization project is underway; completed in 1978, the shares are Texaco 42%, Veba 33%, and Conoco 25%.
[b]50% held by ENI/AGIP.
[c]50% held by Mobil.

Refining capacity by company, 1975[35]

	Thousands of tons	Per cent
BP	20,500	13.3
Shell	18,075	11.7
Esso	25,475	16.6
Veba	30,250	19.7
Other	59,560	38.7
Total	153,860	100.0

Appendix 6. Market Shares

Approximate shares in Germany, 1975[36]

	Petrol (per cent)	Petrol stations[37]
ARAL	25	24.1
Shell	16	11.3
Esso	15	11.9
BP	10	10.6
Texaco	9	10.5
Chevron	2	NA
Conoco	2	NA
FANAL	NA	3.1
Independents including 'white pumps' (i.e. discount stations)	21	NA
	100	

NA = not available.

Veba

	Thousands of tons[a]	Percentage of market
Share of Veba in German sales of:[38]		
Petrol	3,370	17.1
Middle distillate	11,180	20.3
Fuel oil	5,562	24.8

[a]Net of imports and exports.

Appendix 7. German Domestic Oil Production[39]

Production of oil, 1975

		Thousands of tons
Gewerkschaft Elwerath Gewerkschaft Brigitta	} 50% Shell, 50% Esso	1,675
Deutsche Texaco		1,125
Wintershall		822
Mobil		779
Preussag		559
Deutsche Schachtbau		476
Deilmann AG		203
H.v. Rauterkranz		60
Gewerkschaft Deutz		2
Gewerkschaft Norddeutschland (BP)		0.8

Estimated 'windfall profits', 1975[40]

	Millions of DM	Percentage of profit from gas
Esso	250–380	80
Shell	250–380	80
Mobil	212–322	78
Wintershall	121–183	60
Texaco	83–127	18
Preussag	50–78	28
Deutsche Schachtbau	31–50	7
Deilmann AG	24–36	48
H.v. Rauterkranz	12–17	68
Gewerkschaft Norddeutschland (BP)	3–3	100
Gewerkschaft Deutz	1.2–1.7	95

Appendix 8. Numbers of Firms and Market Share in German Oil Industry[41]

Number of firms

	1955	1960	1965	1970	1975
Majors[a]	4	4	5	7	6
Other foreign firms	–	3	4	7	6
German firms	9	6	7	5	4
Small or family firms	6	4	2	–	–
	19	17	18	19	16

Relative market share (per cent)

	1955	1960	1965	1970	1975
Majors[a]	41.5	53.2	55.5	62.8	59.9
Other foreign firms	–	7.9	5.9	13.0	10.3
German firms	55.4	37.0	37.9	24.2	29.8
Small or family firms	3.1	1.9	0.7	–	–
	100.0	100.0	100.0	100.0	100.0

[a]Gulf no longer operates in Germany.

Appendix 9. Sales and After Tax Earnings by Division[42]

Sales

	1975		1974	
	Sales (millions of DM)	Per cent	Sales (millions of DM)	Per cent
Electricity	4,488	18.1	3,933	15.7
Oil, gas, chemicals	8,464	34.1	9,657	38.4
Glass	501	2.0	509	2.0
Other	379	1.5	300	1.2
	13,832	55.7	14,399	57.3
Trade	9,328	37.5	9,236	36.7
Other services	1,688	6.8	1,504	6.0
	11,016	44.3	10,740	42.7
Total	24,848	100.0	25,139	100.0

After tax earnings

	1975 (millions of DM)	1974 (millions of DM)
Electricity	267	199
Oil, gas, chemicals	−179	17
Glass	−15	2
Trade and other	55	60
	128	278

Appendix 10. Profitability Comparisons[43]

Company	Turnover (billions of DM)			Net profit[a] (millions of DM)			Investment in fixed assets (millions of DM)			Personnel		
	1973	1974	1975	1973	1974	1975	1973	1974	1975	1973	1974	1975
Esso AG	8.4	11.0	11.2	292.7	371.0	28.8	279.0	205.7	141.4	4,536	4,445	4,319
Shell AG	7.9	11.2	10.8	162.0	156.1	100.1	352.7	302.2	285.9	5,436	5,128	4,894
Mobil AG	2.6	3.9		197.0	180.0		253.1	540.4		2,283	2,322	
Texaco AG	3.9	5.9	5.2	37.4	36.9	4.6	113.6	93.6	79.1	6,736	6,628	6,293
Veba		25.1	24.8		311[b]	259		391[d]	237		12,709[c]	13,468

[a]Net profit = Bilanz gewinn. (N.B. These figures may not be directly comparable without further research and are intended to serve only as guidelines.)
[b]These figures are profits after taxes – not group profit; excluding Gelsenberg.
[c]These figures include those employed in the chemical side of Veba.
[d]Investments in crude oil refining/chemicals only.

Appendix 11. Veba Group Structure[44]

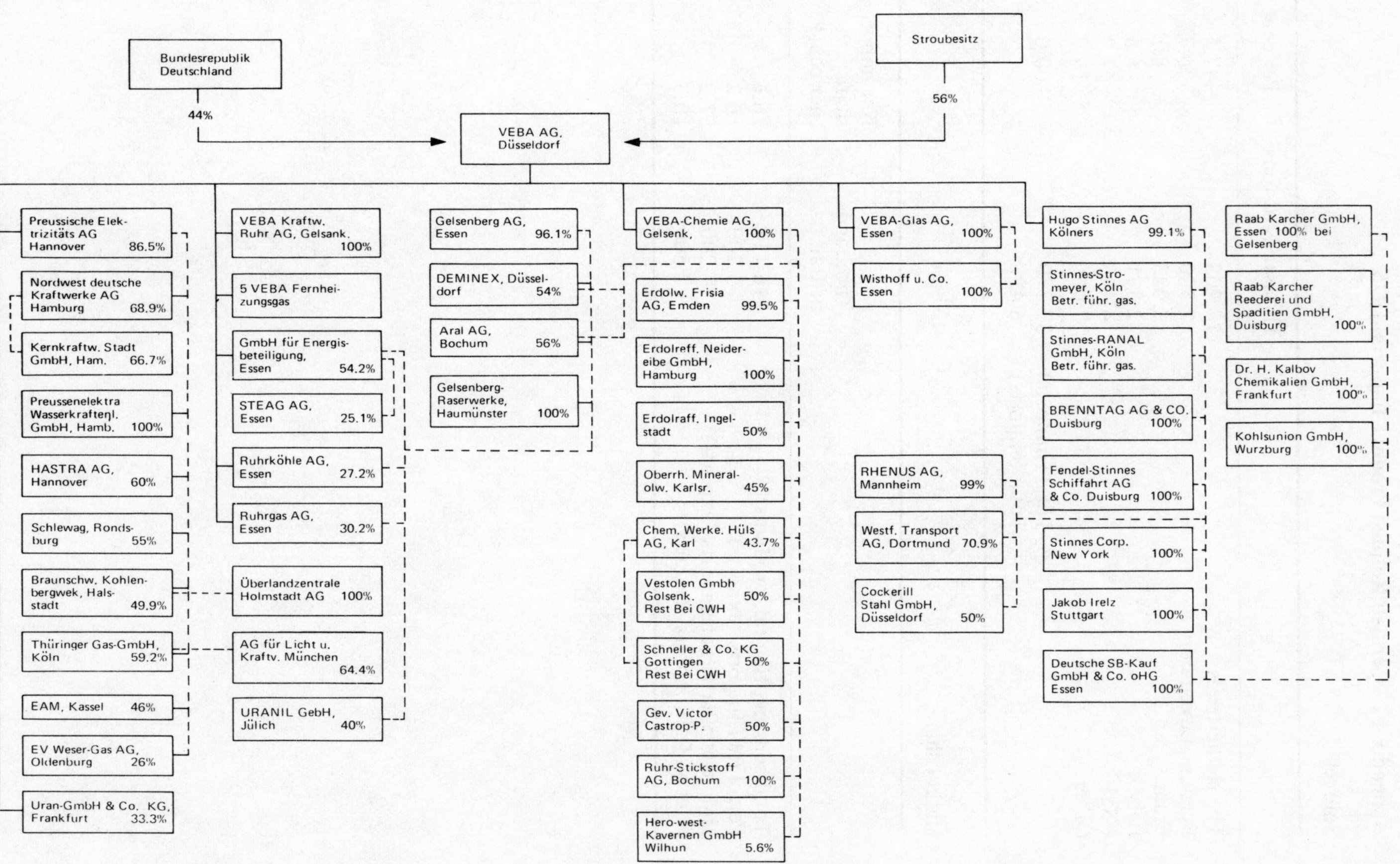

Appendix 12. German Refined Product Imports, 1975

Sources[45]

	Thousands of tons	Per cent
The Netherlands	17,703	47.5
Belgium/Luxemburg	2,500	6.7
France	3,023	8.1
Italy	2,531	6.8
USSR	4,447	11.9
Other	7,066	19.0
Total (imports)	37,270	100.0
Total (exports)	6,436	

Market share

	Thousands of tons	Net import share (per cent)
Total consumption (all products)[46]	116,118	26.6
Petrol (net imports–exports)[47]	3,200	16.2
Middle distillate (net imports–exports)[41]	17,200	31.2
Fuel oil (net imports–exports)[41]	2,300	10.3

References

1. *Report by the Veba Company*, 1975.
2. *Gelsenberg AG*, Verlag Hoppenstedt & Co., Darmstadt, 1969.
3. Gelsenberg AG, *Annual Report*, 1973; Veba AG, *Annual Report*, 1973.
4. Anthony Sampson, *The Seven Sisters*, Hodder & Stoughton, London, 1975.
5. Georg H. Düster, 'Germany', in *Big Business and the State*, Raymond Vernon (Ed.), Harvard University Press, Cambridge, 1974.
6. Georg H. Düster, 'Germany', in *Big Business and the State*, Raymond Vernon (Ed.), Harvard University Press, Cambridge, 1974.
7. Georg H. Düster, 'Germany', in *Big Business and the State*, Raymond Vernon (Ed.), Harvard University Press, Cambridge, 1974.
8. Horst Mendershausen, *Coping with the Oil Crisis*, Johns Hopkins University Press, Baltimore, 1976.
9. Horst Mendershausen, *Coping with the Oil Crisis*, Johns Hopkins University Press, Baltimore, 1976, p. 24.
10. Von Herbert Lötgers, *Sonderdruck aus Zeitschrift 'Erzmetall'*, Band 25 (1972) Heft 55255–259, Die Deutsche Erdölversorgungsgesellschaft, p. 257, and Deminex: Ziele und Aufgaben im Rahmen der Deutschen Rohölversorgung – 'Die Mineralölindustrie in der BRD', Frankfurt, 1972.
11. Gelsenberg AG, *Annual Report*, 1974.
12. Von Herbert Lötgers, *Sonderdruck aus Zeitschrift 'Erzmetall'*, Band 25 (1972), Heft 55255–259, Die Deutsche Erdölversorgungsgesellschaft, p. 2.
13. 'The Philosophy of Exploration: West Germany's Deminex Explains Its Objectives', *Petroleum Times*, November 28, 1975.
14. Von Herbert Lötgers, *Sonderdruck aus Zeitschrift 'Erzmetall'*, Band 25 (1972), Heft 55255–259, Die Deutsche Erdölversorgungsgesellschaft, p. 2.
15. *Der Spiegel*, November 11, 1974.
16. Veba, *Preliminary Report*, 1975.
17. *Jahresbericht 1974*, Mineralölwirtschaftsverband e.V. Arbeitsgemeinschaft Erdöl-Gewinnung und-Verarbeitung (MWV/AEV), p. 17.
18. *Jahresbericht 1974*, Mineralölwirtschaftsverband e.V. Arbeitsgemeinschaft Erdöl-Gewinnung und-Verarbeitung (MWV/AEV), p. 17.
19. ARAL, *Annual Report*, 1974.
20. MWV/AEV, p. 14.
21. MWV/AEV, p. 9.
22. Veba Mineralöl, '1975 Market Share', Press Conference, March 12, 1976.
23. Private communication to the author, Spring 1979.
24. BP, Press Release, London, June 16, 1978.
25. *Wall Street Journal*, October 3, 1978.
26. BP, Press Release, London, March 5, 1979.
27. Private communication to the author, Spring 1979.
28. *Der Spiegel*, Nos. 11 and 16, 1979.
29. Veba AG and Gelsenberg AG, *Annual Report*, 1973.
30. Esso (Deutschland) AG, *Annual Report*, 1975, p. 5.
31. *Deminex Activity Report*, No. 2.
32. *Deminex Activity Report*, No. 2.
33. MWV/AEV, *Jahresbericht 1975*, Hamburg, 1976, p. T-10.
34. Veba, *Preliminary Report*, 1975.
35. Walter Monig *et al.*, *Konzentration und Wettbewerb in der Energiewirtschaft*, R. Oldenburg Verlag, München, 1977, p. 287.
36. Private communication to the author, 1976.
37. *Konzentration*, Tabelle II 87b.
38. Veba Mineralöl. '1975 Market Share', Press Conference, March 12, 1976.

39. *Konzentration*, Tabelle II 61.
40. *Konzentration*, Tabelle II 64.
41. *Konzentration*, Tabelle II 69.
42. Veba, *Annual Report* (in German), 1975, pp. 13–14.
43. *Annual Reports.*
44. *Konzentration*, p. 224.
45. MWV/AEV, 1975, p. T-25.
46. MWV/AEV, 1975, p. T-9.
47. Veba Mineralöl.

CHAPTER 7

British National Oil Corporation

British National Oil Corporation (BNOC) was created by an Act of Parliament on January 1, 1976. Its incorporation was the culmination of a series of developments in the formulation of a policy for the exploitation of Britain's oil resources, which had spanned three governments since the first round of exploration licenses for the British sector of the North Sea in 1964. Oil in Britain was viewed both from the point of view of benefit to the economy in general and to alleviate the chronic balance of payments deficits, the latter becoming even more crucial with the growing economic difficulties of the early seventies, culminating in the energy crisis when it became of paramount importance. Balance of payments deficits contributed to the worsening of inflation, unemployment, growing international indebtedness, and the weakening of the pound.

Britain is the only large industrialized country in the West with a prospect of self-sufficiency in energy supplies in the immediate future. Moreover, by 1980, British oil production will match that of Venezuela and Nigeria, fourth and fifth producers in OPEC, and Britain will be the world's largest producer of low sulfur crude. In addition, North Sea oil may provide a solution – or create additional problems – to regional issues, particularly, but not only, Scotland. These problems are general in nature but affect the Labour Party particularly.

Among other considerations British North Sea oil policy is an illustration not only of a change in governmental philosophies, from the Conservatives' *laissez faire* to Labour's more interventionist policy and back again, but also how governmental thinking has been affected by external considerations, namely the quadrupling of crude oil prices in the early seventies and the large price increases of 1979 and the consequent change in the profitability of North Sea oil and gas.

North Sea Background

In 1958 an Esso/Shell consortium discovered the Groningen gasfield in Holland and the effects of this discovery were swift and impressive. Reserves estimated at 1,900 billion m^3 were in excess of foreseeable Dutch domestic needs, and by 1970 natural gas from this field accounted for 29 per cent of Holland's primary energy requirements. Exports of natural gas stood at 36 per cent of total production. The magnitude of the Groningen field spurred the oil companies into exploration on the Continental Shelf area of the North Sea. British Petroleum (BP) made the first discovery of gas in the southern part of the British sector in 1965, and by 1967 enough gas had been discovered, and future prospects were considered good enough, for the decision to be taken to convert progressively the whole of the interconnected gas demand in Britain to natural gas supply.[1]

In 1969 the consortium headed by Phillips Petroleum made the first discovery of oil in large quantities when it struck the Ekofisk field in Norwegian waters. This discovery marked the beginning of the second phase of North Sea exploration activity, since it confirmed to the oil companies that the oil which they had been hoping for really did exist in large quantities. The Ekofisk find was followed by the Montrose field discovery and, later, by BP's discovery of the Forties field in the British sector. The discovery of these fields banished all remaining doubt as to the commercial viability of North Sea oil reserves, and was the signal for a sharply increased exploration activity. Between the beginning of 1972 and the end of 1976, fourteen significant fields were discovered with total reserves, in 1976, at 17 billion barrels of proven reserves.[2] The first oil came ashore in 1975 by tanker from Argyll. In 1977 nine fields were on stream; the remaining five were under development. Two pipelines to shore were already in operation and 1977 production was 35–45 million tons and about 100 million tons in 1980. Yield to the Exchequer from royalty petroleum and corporation taxes was likely to be around £5.5 billion by 1980 and about £3.5 billion per year thereafter, both in 1975 prices.

The Evolution of British Government Policy

When faced with the production and exploitation of North Sea gas, the British government had a relatively easy task in formulating a policy for the following main reasons. The gas industry had been nationalized in 1949, and had the technical infrastructure to deal with production. There were relatively few private companies involved in the offshore operation and production was for domestic consumption only, because of the large size of the domestic market. There was no question of exports, as in the case of Holland. (However at the end of 1977, exports were being considered for the early 1980s. These would not be net exports; Norwegian gas would be imported into Britain and British gas exported to the continent.) There existed a close integration between the gas-supply industry and the marketing, sales, and services organizations. Natural gas first augmented and then displaced town (coal) gas and oil-based gas.

However, the problem facing the government on the issue of oil was far less clearly delineated.[3] Licenses had been awarded first in 1964, then in a second round in 1965, and the main government effort was directed toward encouraging the oil companies to explore offshore Britain, not toward tackling the policies which would be needed once oil was found in any quantity. This was understandable in the context of the times. The overwhelming need of a country with few raw materials, but with a considerable industrial consumption of oil and a balance of payments problem, was to find out what kind of resources did lie off-shore, and to find them at the minimum cost to the state.[4] At that time, the government did not have the capability to deal with the problem on its own, or it perceived that it did not have the capability, and thought it necessary to rely more on the oil companies. The North Sea was a high-risk area, to be explored with a then-unknown technology and in a period of oil surplus and low prices. Middle East oil was still plentiful and cheap.

Against this background, the licensing policy executed by the Ministry of Power, adopted by the Conservative government of 1964 and its Labour successor of 1964–70, had much to recommend it. (Licensing rounds took place in 1964, 1965, 1970, and 1971–72.) By making rentals and the initial financial terms relatively cheap,[5] and concentrating on the work program commitments as the competitive element in license applications, it made the area as attractive as possible to outside investment, while ensuring that drilling was actually carried out. At the same time, the Ministry of Power adopted a discretionary system of allocation, rather than an open bidding system; out of the 2,655 blocks offered and 245 licenses awarded only 15 were not allocated at the discretion of the Minister. This, together with the statement that preference would be given to companies with the necessary financial resources, with exploration experience, and with an established record of contribution to the British economy, favored British-based companies in a way in which a bidding system could not have done. Preference was also given to companies that had established refining and marketing operations in Britain. Approximately 33 per cent of the acreage went to British, or part-British companies, 45 per cent to American companies, and the rest to others. Most of the American companies had, and have, substantial British operations. The onus of the early exploration rested with the major oil companies, and with BP, Shell, and Esso in particular. The majors were the only companies which were technically and financially competent to cope with the arduous conditions of the North Sea and there did not seem to be any particular reason for a government agency to become involved.

The early years of the seventies brought events which revealed the potential weakness of this North Sea policy, from the point of view of the government. At the same time as OPEC was demonstrating its effectiveness in asserting its control over the ownership, production, and price of crude oil in the period 1971–73, discoveries were being made in the North Sea. Thus, while most European industrialized countries were facing a period of supply and price pressure from OPEC, Britain had the prospect of becoming self-sufficient in oil by 1981. The political atmosphere changed from one in which the aim was that of strongly encouraging exploration to one of rapidly trying to work out the consequences of being a potential major oil producer.[6] The main weaknesses of the North Sea policy as far as the government was concerned were pointed out in a speech made by Lord Balogh in the House of Lords in 1975.[7]

> There were no controls over the rate of production/depletion. There was no obligation for licensees to continue exploring after the first six years. (A license was granted for a period of 46 years.) There was no adequate provision for the Government to receive financial information from licensees. There was no clause in the tax provisions to

> prevent alleged excess profits deriving from the increase in oil values following OPEC's price rises although the oil companies claimed that the OPEC price rise was unforeseen and they could not have realized the windfall profits. (See subsequent discussion.) There was no control over the flaring of gases from oil wells. And finally there was insufficient linkage with British industry to ensure that it was prepared to take a major share of the opportunities provided by the North Sea. (What is meant by insufficient is, of course, a matter of dispute.)

Early in 1973 Harold Lever, then in Opposition and Chairman of the Public Accounts Committee (PAC) launched an attack[8] on the tax measures for the North Sea, and recommended actions to improve substantially the tax yield and secure better licensing terms for the Exchequer. It was while Ministers and officials were trying to defend themselves against the PAC charges that the quadrupled price of OPEC oil raised the prospect of potential windfall profits from the North Sea.[9] In response to this situation, the Conservative government began to work on a tax aimed at preventing excessive profits,[10] and by the beginning of 1974, i.e. before the General Election of February that year, a movement toward a radical reappraisal of North Sea oil policy was gathering momentum.

The companies maintained that they had never expected to be the recipients of windfall profits.[11]

> 'I should at the outset caution against belief that, just because OPEC countries have increased the price of competitive supply, profits from North Sea oil will automatically be vast. The level of Government "take" from North Sea profits is under constant scrutiny, and it would be unreasonable to think that this level will not be influenced by developments in OPEC countries. Profits to stimulate exploration and development will, I am sure, be made – but I would curb some of the more extravagant optimism.'

Relations between the British Government and the British and Part-British Owned Majors

From the earliest days of World War I, successive governments had encouraged BP and, to a lesser extent, Shell in their capacities as guardians and promoters of the nation's interests in the face of American dominance of the oil industry. Although the government had two appointments to the Board of BP, with powers of veto, it had been agreed with Churchill in 1914 (and this agreement had been honored ever since) that these powers would only be exercised on questions of foreign or military policy, or on matters bearing directly on Admiralty contracts. Although the Foreign Office had been critical of BP's chairman, Sir William Fraser, for his handling of the Iranian crisis of 1951–54,[12] this had been the only major disagreement, and there had never been any real reason to suppose that the interests of BP and the British government were anything but roughly parallel.

Shell, though to a somewhat lesser extent, enjoyed the same spirit of alliance with the government as BP. (Shell is 60 per cent Dutch owned.)

However, by the end of 1973, relations between the government and BP had become distinctly strained. Not only was BP a large potential North Sea producer, but it had insisted on honoring its contracts with foreign customers at the time of the Arab oil embargo, and not diverting supplies from these customers to Britain. This had provoked a bitter argument between Sir Eric Drake, BP's chairman at that time, and Edward Heath, the Prime Minister. The government began to look more closely at the question of control over BP, but the Foreign Office insisted that intervention would jeopardize BP's relations with governments of other countries, particularly the United States, where BP had large interests in Alaska. Instead, a solution in an altogether different guise began to emerge. In order to gain control over the North Sea while appeasing the Scottish Nationalists, who were threatening to secede from the United Kingdom with 'their oil' (sic),[13] the government could consider participation.

The establishment of a national oil company was embodied in the Labour Party election manifesto for the February 1974 general election. It was stated that, if elected, they would:[14]

> . . . Take majority participation in all future oil licenses and negotiate to achieve majority State participation in existing licenses. Set up a British National Oil Corporation . . . Take new powers to control the pace of depletion, pipelines, exploration and development.

After their victory in that election, the Labour Party pursued the goals contained in this pledge by moving forward on three fronts, tougher taxation, tighter administrative control, and 'National' control.

Labour Party Initiatives

Taxation

The Oil Taxation Act of August 8, 1975, introduced the Petroleum Revenue Tax (PRT) designed to cream off the potentially high North Sea profits because of the quadrupling of oil prices set by OPEC. It gave the Inland Revenue the power to challenge transfer prices where these prices applied to transactions between one tax resident company and another; this point applied almost entirely to BP and Shell. And it allowed only £50 million for tax purposes of the large, accumulated losses by reason of high intercompany prices which were charged up to July 1974. This provision, again, applied almost entirely to BP and Shell since the US companies were not tax residents in the United Kingdom. (They, of course, payed UK tax.) Finally, the Act established a 'ring fence' around the North Sea such that any downstream losses could not be offset against producing profits. On non-North Sea oil, tax resident companies obtained relief in the United Kingdom for tax paid abroad. However, the 'ring fence' allowed offshore losses to be offset against on-shore profits which was one explanation of the interest of the many nonoil companies in the various exploration consortia.

Initially, the Public Accounts Committee proposed a single rate Petroleum Revenue Tax (PRT) based on cash flow. The main advantage of this idea was its simplicity. However, it was realized that it would place an intolerable burden on the smaller, higher-cost fields. The PRT was then revised.[15] (See Appendix 1 for specifics.)

The base rate of the PRT was fixed at 45 per cent and was allowable before liability to Corporation Tax. The government had promised to review this rate and to offer relief from royalties, on a discretionary basis, to fields in difficulty. Under normal operating conditions, PRT, plus royalties (12.5 per cent of well-head value of oil), plus Corporation Tax (currently 52 per cent) should ensure the government of a tax take in the region of 70 per cent.[16]

Tighter Administrative Control

One of the strongest criticisms which the Labour government made of the Conservative North Sea policy was the extent to which there was insufficient control over the way in which offshore operations were being carried out. The Petroleum and Submarine Pipelines Act (passed on November 12, 1975) contained provisions which empowered the Minister of Energy to impose depletion rates on fields,[17] enforce common carrier provisions on pipelines,[18] and to sanction both the timing and direction of new development expenditure.[19] These powers in turn were tempered with a number of provisions, giving the companies the right of appeal, ensuring that decisions did not upset commercial considerations too far, and provided some limitations on the depletion and pipeline powers of the Minister. Some of these provisions were further amended in the Energy Act of 1976.

National Control

The most important new development represented by the Labour Party policy was the establishment of a national corporation to represent actively the nation's interest in the exploitation of North Sea oil. The main reasons that may be advanced for a government agency or oil company being set up were as follows. (These were not necessarily the reasons given by the government and, in turn, the government may have had additional reasons as well. To some extent the British were influenced by the establishment by the Norwegians of STATOIL in 1972.)

The first reason was ideology, that valuable resources, oil and gas, should be used for the 'public's benefit'. Second, it was considered necessary to monitor an important indigenous resource. The government did not have the 'know-how' to explore and develop offshore fields and it was claimed that participation would allow it to gain this expertise. Then, without some control, through participation, there would be no way to ascertain costs and prices, i.e. true profits, and the whole tax policy would collapse. Coal, gas, and nuclear power were publicly owned; by the same token so should oil be, and the creation of a government oil company would be an alternative to nationalization. (This point would not have necessarily been acceptable to the Conservatives.) Finally BNOC would not have to comply with Common Market (EEC) energy policy if and when such policy emerged, while the majors, which operated in some or all member states, would be subject to much greater pressure from the EEC. Theoretically, EEC decisions applied to all companies but for practical purposes the EEC had been able to do little about governments' assistance to their own enterprises.

BNOC had been set up from the start as a separate government-controlled corporation. The broad range of powers vested in the company by the Petroleum and Submarine Pipelines Act left it in no doubt that the corporation was in a position to build up a competitive competence in all the operations of a major oil company, at least on a national scale. Whether or not it would have the ability to compete internationally was to be seen. It had a brief covering exploration and production, not only in the North Sea but in any part of the world.[20] Nor was it intended that its activities be restricted to the exploration for and production of oil. It was empowered to operate tankers, build refineries, and sell finished products.[21] In sum, it could become a large international integrated oil company with refineries and a marketing network outside the United Kingdom as well as inside.

BNOC

The results of the PAC review and the consequent 1974 White Paper[22] set out a series of objectives which included the setting up of BNOC and the arrangement for state participation of both existing and future licenses. The powers, activities, duties, and financing of the corporation[23] were as follows.

BNOC was to explore for and produce petroleum, to transport and refine petroleum, to store, distribute, buy, and sell petroleum and derivatives, to take over HMG's participation interest in UK licenses, to carry out consultancy, research, and training, and to build, hire, and operate refineries, pipelines, and tankers. BNOC was empowered to set up or acquire subsidiaries, and to give loans and guarantees (restricted to oil companies). The corporation was to act in accordance with plans and budgets agreed with the Secretary of State, to give advice to the Secretary of State on petroleum matters, and to take over the administration of the government's defense pipeline and storage system, as required. BNOC had powers to borrow from the government in sterling, or from others in either sterling or foreign currencies, with borrowing and guarantees limited to £600 million, increased to £900 million by order of the House of Commons.

In June 1977 BNOC took its first plunge in the international capital markets, raising $825 million over eight years from twelve banks, both domestic dollars and Euro-dollars,[24] and the acceptance of BNOC as a corporate borrower gave it added credibility as an entity should subsequent governments ever wish to dismantle it. The Secretary of State was empowered to make loans to BNOC and

BNOC was subject to all taxes except PRT, but there was an obligation on it to calculate the amount of PRT that would have been payable in the absence of such exemption.

National Oil Account

Under Part V of the Petroleum and Submarine Pipelines Act, a National Oil Account (NOA) was set up to collect all royalties from the North Sea oil production. Royalties were calculated at 12.5 per cent of the well-head value of the oil (that is f.o.b. minus cost of transmission from field to shore, or the net of cost of conveying and treating oil). By assuming an f.o.b. price of $13.70 per barrel, a transport cost of $0.20 per barrel, and a dollar to pound exchange rate of 1.80, the calculation given in Table 1 shows the potential size of the NOA. (By 1980, the f.o.b. price of North Sea oil was about $30 per barrel.)

To the amount contributed by royalties would be added any profits (or losses) which BNOC might make. In fact, BNOC's gross revenues were to go into the NOA and all borrowings would go through the NOA, which would be under the control of the Secretary of State for Energy.[25] It was his responsibility to declare whether or not the account was in surplus; the surplus would go to the Treasury. This control gave the Secretary of State discretionary powers as he could direct expenditure from the account, including the option to go abroad or downstream.

Initially, at least, BNOC would be able to count on government support for its spending. But by the time the government (and Parliament) were willing to scrutinize BNOC's investment plans, the company might have an adequate cash flow; and BNOC would be different from other British nationalized industries. The latter, because of their generally poor commercial performance, received public scrutiny. BNOC would produce net revenue and would therefore receive less public examination.

Table 1. Amounts collected in the National Oil Account

	1975	1976	1977	1978	1979	1980	1981	1982
Production volume (thousand barrels per day)	23	340	861	1,334	1,852	2,587	3,017	3,250
Well-head value at $13.50 per barrel (millions of £)	62	916	2,319	3,593	4,988	6,968	8,127	8,754
Royalties at 12.5% of well-head volume (millions of £)	8	115	290	449	623	871	1,016	1,094

Source of production figures: Wood Mackenzie.

Participation

The Labour Party had stated in its manifesto for the 1974 general election that it would pursue a policy of participation in the North Sea. The term 'participation' was not clarified and considerable confusion arose as to whether it implied equity participation and to what extent oil company control over exploration and production would be affected. By 1976 many doubts had been resolved through a series of statements, and a fledgling policy had begun to emerge. BNOC would gain a right to 51 per cent of oil produced from, rather than an equity share in, all commercial fields, and participation would be 'voluntary' for those blocks already licensed.

The government repeatedly assured the oil companies under the first four licensing rounds that participation would not involve any financial loss or gain to the company concerned. It was never made clear what the operational meaning of that statement was. ('If it is a "financially no better and no worse off" situation, why do it,' said an oilman.) The government indicated that the market price was to be paid for the oil taken, but the companies claimed that for reasons of diversification of supplies they wanted the crude itself. However, a main condition for the allocation of blocks in the 1977 licensing round was that BNOC had 51 per cent interest in any block.

In that sense, the government's insistence that there was no compulsion involved in accepting participation was hard to comprehend. Since the Ministry of Energy had not only the power to grant licenses but also to regulate the existing licenses, it was hard to visualize a situation where available pressure was not used. The Secretary of State for Energy had suggested that it was in the companies' interest to allow government participation if they expected to extend their operations on the Continental Shelf (CS) and benefit from future discretionary awards of licenses.[26] Pressure was brought on all companies, including BP, Shell, and Esso, and all major UKCS operating companies had agreed to participation at least under outline agreements by 1977. (For market and refinery shares see Appendix 2.) Some companies accepting participation were obliged to grant BNOC the right to buy 51 per cent of their oil at market prices.

The proposed terms of the fifth (1977) licensing round, published in a consultative document on May 27, 1976,[27] also contained further clarification of the government's evolving oil policy. BNOC would pay its share of all exploration work. This could involve the corporation in a contribution of over £100 million but was minor in comparison with development costs. Benn, who succeeded Varley as Secretary of State for Energy on June 10, 1975, stated that by contributing directly towards the search for oil, the corporation would be able to exert greater influence as well as gain greater knowledge and experience. ('If we were not in there in a real sense, we would have significantly less control over development.')[28] Second, after completion of the exploration phase, BNOC would have the right to opt out. Its co-licensees would then proceed with the development on their own, having complete jurisdiction over the oil produced and the assets used. However, BNOC had the right of re-entering later by paying interest.

Finally, relinquishment of acreage in licenses was more onerous, requiring a surrender of two-thirds after seven years. This compared with the previous requirements of one-half relinquishment after six years.

BP, BNOC, and the Secretary of State for Energy signed a Heads of Agreement in July 1976.[29] Accordingly BP not only agreed to BNOC participation but undertook the training of BNOC staff in its (UK) refining and marketing operations. This latter provision was as important as BNOC's enlarged access to crude as BNOC had, so far, been short of expertise. There were also provisions for downstream cooperation between BP and BNOC. BNOC stated that it did not intend to undertake independent marketing operations before 1980 but the implication was, however, that after 1980 with both crude and a well-trained staff it would enter downstream operations. The BP–BNOC agreement also provided for BP to supply oil to BNOC from other worldwide sources in return for sale-back of UK Continental Shelf oil. The BP–BNOC settlement put considerable pressure on Shell and Esso and the other North Sea oil operators, and the chairman of Shell Transport and Trading said that 'now that BNOC is in existence we may as well try to keep it on the right lines.'[30] He went on to express his apprehension that any company, like BNOC, which had large quantities of crude available was a 'very unstable element in the market place.'

In February 1977 Shell, Esso, Mobil and Texaco signed Heads of Agreements with BNOC and the government under a formula giving them buy-back rights to their oil production in the UK offshore, with consultation process limited to national oil policy and availability of supplies. Shell and Esso also agreed to let BNOC staff be trained not only in refining and marketing but also in their planning operations. By early 1977, all companies operating in the British sector of the North Sea had signed participation agreements with the Secretary of State for Energy and BNOC. Amoco was the last company to sign; it did so in April 1977. The government was clearly pleased with the oil company response in the fifth round as altogether sixty-five companies including all the majors had laid claim to 9,000 km^2 around the United Kingdom.[31] The fifth round took place in February 1977 after virtually all Heads of Agreements had been signed with the private companies.

The Corporation

The decision to set up BNOC, and the powers vested in it in the Petroleum and Submarine Pipelines Act, provoked sharply divided opinions as to the nature and effectiveness of the corporation. Lord Balogh, long-time advisor of Harold Wilson and the chief architect of the act, saw BNOC as the guardian of the British people's rights and a foundation for Britain's economic revival, while Patrick Jenkin, the former Conservative Shadow Minister for oil, described it as 'faceless, flabby and purposeless.'[32] In between were the oil companies themselves, angry sometimes at what they considered a betrayal of the confidence embodied in the terms of the licenses as originally granted, and at other times simply confused by the looseness of important issues such as participation, but basically resigned to the existence of BNOC. In the United Kingdom, the major companies enjoyed their best government and public relations of all the large OECD countries.[33] Simultaneously, a welcoming (to BNOC) attitude arose saying, in essence, 'we shall beat you in the market place'. After all, companies had learned to live with national oil companies, with various degrees of reluctance or enthusiasm, both in the producing and consuming countries.

Now that it had been created, would BNOC remain a 'robot',[34] tightly controlled by the Department of Energy, or would it, through an extension of its operations into refining and marketing, grow to resemble its natural competitors in these areas, the major oil companies? Much depended on the character of the government in power, and the personalities involved in administering BNOC. The Labour government appeared intent on using BNOC as an instrument to obtain control over North Sea oil reserves. It had pushed forward on the basis of participation and aimed at achieving 51 per cent control through the option to purchase production at market prices. Labour policy, in the short term at any rate, was also colored by Benn's own special views on national control of the North Sea; under his influence, BNOC would surely be closely controlled by the Department of Energy. If this directive trend was maintained, BNOC management would be concerned mainly with the implementation of the policies of Benn's department, rather than with formulating and applying their own decisions. However, this did not take into account the possibility that BNOC might develop as the center of decision making once revenues started accumulating. Although BNOC might be dependent on the Department of Energy at present, a solid financial base could be the beginning of a much more independent course of action.

If the Conservatives were to come to power, it seemed unlikely that BNOC would change very much, although modifications could be expected. They would establish a United Kingdom Oil Conservation Authority (UKOCA) and Mrs Thatcher had implied that BNOC would be dismantled 'as far as possible'. However, the past record of the Conservative Party in decreasing state ownership was not at all clear-cut. UKOCA would be a small and highly expert authority under the control of the Secretary of State for Energy, which would control the timing of oil company investment in production facilities, rather than the actual rate of production itself. More emphasis would be placed on taxation to provide revenue than on participation.[35] (Government policy after the Conservative victory is discussed in a subsequent section.)

Under its terms of reference, BNOC was permitted to extend its activities into refining, shipping, and marketing, as well as exploration and production anywhere in the world. Marketing, being a less profitable phase of the industry, could be the last phase for BNOC to enter. It seemed unlikely that it would not be involved in marketing operations and, as a matter of fact, might be forced into it earlier than planned by the default of Burmah[36] and other smaller cash-starved marketers. Moreover, about 25 per cent of oil consumed in the United Kingdom was by nationalized bodies, ranging from industries to the National Health Service and the Armed Forces. If local authorities and the rest of the public sector entities were included, the figure was about 40 per cent. The government could set guidelines which could force these entities to buy from BNOC. Such guidelines would be in contravention of EEC guidelines but this would not be the first or last time that these guidelines would have been violated. These implications had been denied by the government. It was likely at this stage that the government would follow the German example in its arms-length dealings with Veba, rather than the French and Italian governments' strongly interventionist policies favoring, respectively, Elf-Aquitaine and ENI. For the present, the actual future course of British policies is a matter of conjecture.[37]

That marketing would be the natural extension of any BNOC initiative in refining seemed likely in view of the Labour government's apparent intention to put pressure on all North Sea producers to refine up to two-thirds of North Sea crude in Britain. This would be a political decision and it was unlikely to persist against balance of payments and other commercial considerations. Because of the low sulfur content of North Sea oil, the demand for North Sea oil in the United Kingdom was limited. What was more likely was that the United Kingdom would collect the benefit of a sulfur premium in export markets for premium crudes. Alternatively, the UK refiners could export products. However, with the present European-wide excess in refining capacity, this latter course of action faced considerable obstacles. The companies so far had resisted, saying that they would maximize their profits if only 35 per cent went to British refineries.[38] If the government insisted on 66 per cent, one way for it to force it through would be for BNOC to exercise its rights to build refineries, and to refine the balance of 25 per cent of its 51 per cent option on production. Under the hypothesis that 51 per cent participation was achieved by 1980, this would mean a requirement in that year for a BNOC refining capacity of 31 million tons per annum on the basis of the latest estimates of production. There were two problems, however. The first was that far from needing extra refining capacity, the United Kingdom had unused capacity. (In 1975, capacity was 147 million tons, while consumption was 89.5 million tons.)[39]

The second problem was that the UK's present refining programs were geared to Middle East medium crudes, and used very little light crude. In 1976, estimates were for a total light crude consumption of 25 million tons out of a total UK consumption of 90 million tons. Most North Sea crude was very light; unless the government pressed ahead and built a BNOC refinery network specially programmed for light crudes (unwise with the present capacity underutilized) the United Kingdom would appear to be heading for a surplus of light. The UK refineries could handle North Sea crudes. There was a dichotomy of views here between maximizing profits on the one hand and maximizing fiscal revenue on the other. BP, Shell, Esso, and the other majors claimed that they could not maximize profits if they had to refine 66 per cent of North Sea crude in the United Kingdom. Maximization of governmental fiscal revenue might, or might not be, maximized under the 66 per cent guidelines.

A likely solution under these circumstances was that BNOC would be drawn into the international exchange market, selling North Sea light to those countries such as the United States which had a strong demand for light, low sulfur oil in return for medium and thus reap the premium. A move of this nature might have an influential effect on BNOC's corporate identity as it would bring with it a palpable need for flexible thinking and rapid decision making. Fulfillment of these needs would probably draw out BNOC from the blanket cover of the Department of Energy, and give it a more international corporate profile and the beginning of a nationally owned and multinationally behaving oil company. BNOC was already developing plans for marketing substantial quantities of oil from 1978 onwards. From BNOC's present sources, but excluding participation oil subject to sale-back provisions, BNOC could have access to 30 million tons per year in the early eighties. BNOC might also market oil taken as royalty in kind by the Secretary of State.[40]

The Interim Lessons

Oil was discovered in the Norwegian and British North Sea at approximately the same time. Norwegian government oil policy, so far, appeared to have an impressive record of clarity, consistency, and acquisition of expertise. The approach of successive British governments, so far, could be characterized as *ad hoc*, and the outcome influenced by the principal personalities involved.

This assessment was reached even if one took into consideration the different goals pursued by Norway (relatively slow growth) and the United Kingdom (to reach a net export position without undue delay). It had been for political reasons (to mollify the Scots) that the government had decided to locate BNOC headquarters in Glasgow. One result of this was that initially BNOC had had difficulty in recruiting suitable staff. (BNOC inherited from the National Coal Board a nucleus of staff.) This had changed by late 1976. BNOC had managed to attract high-ranking oil company executives, had acquired the staff of Burmah Oil Development of slightly less than 300, and applications for jobs were coming in at a high rate.[41]

National oil companies have almost invariably been influenced by powerful personalities and BNOC was no exception. The chairman was Lord Kearton who, with his forceful personality and commercial experience as the former chief executive of Courtaulds, had made his reputation by building Courtaulds into a challenger to ICI. In accomplishing this goal, he 'learned the ropes' of government subsidies, protection by tariffs, and of acquiring small and independent companies that may have posed a competitive threat. Together with Lord Kearton one of the two full-time members of the 'Corporation', i.e. the Board, was Ian Clark, 41 in 1980. Prior to his appointment, he was for eight years chief executive of the Shetlands Islands Council. Since the Shetlands represented a most important logistical base for North Sea exploration, he had come into frequent contact with oil companies. Nevertheless, he lacked direct oil company experience. The deputy chairman was Lord Balogh who had been economic advisor to the Cabinet and former Prime Minister Wilson since 1964, the year in which the first offshore exploration license was granted. He had been at Oxford University since 1945 and was a teacher and senior colleague of Harold Wilson. Unlike other British publicly owned corporations, BNOC's Board included two civil servants, one each from the Treasury and the Department of Energy. The remaining members consisted of two bankers, two businessmen, two unionists, and one consultant. Members of the Corporation (Board) were appointed by the Secretary of State.

The then Labour Secretary of State for Energy Benn, was in 1966 Minister of Technology, where he was also responsible for atomic energy. Technology had 'wanted to drag the U.K. into the 20th century'. The Ministry vanished, but Benn had, as of 1977, a second and probably better chance of doing what the government failed to do a dozen or so years ago.

'But the greatest uncertainty of all surrounding the BNOC is just what is it there for?'[42] Was it there to reassure the public? Was it there to take risks and undertake investments that private companies were unwilling to make? What was the 'right' balance between strictly commercial behavior and acting in the public interest? If there was a divergence between private and public interests – as there almost inevitably was – who decided how to resolve this divergence? BNOC or the Ministry of Energy? The outcome would depend, to some measure at least, on the respective sizes and competences of the Ministry's and BNOC's staffs. BNOC's employment in mid-1977 was about 450. At the end of 1976 it was 388; of these, 35 people worked in the oil division of the National Coal Board and 296 for Burmah.[43] In the Department of Energy, 41 officers of principal level and above were involved in North Sea oil and BNOC matters; of these, 18 had 'relevant qualifications or have had recent appropriate North Sea experience'.[44] BNOC and the government had already disagreed on the price BNOC had to pay for the assets of Burmah, BNOC thinking that the price was too high. They had also disagreed on whether or not the BNOC policy of paying for development costs on the basis of 'carried interest' should be continued. The government thought so and BNOC did not. Finally, if BNOC was there to represent a 'window' into the oil industry, some BNOC participation seemed inevitable in both refining and marketing. The BP deal in particular, and to a lesser extent the Esso and Shell deals, had taken care of this problem, at least until 1980. If downstream operations let alone upstream operations outside the United Kingdom were being envisioned, one might legitimately ask: why would BNOC succeed when such established MNCs as, say, Gulf and a couple of others are doing only fairly?

The outcome would, more likely, be determined by personalities, by the kind of government Britain would have, by how long Benn would stay at the Ministry of Energy, and lastly but significantly by the policies and implementation capabilities of the BNOC executives.

BNOC and the Conservative Government

The Conservative government won the election of May 1979 with a considerable margin. Both before and after the election the role of BNOC had been debated with more heat than light. To be sure, after the election some changes have been made by the Conservative government; however, the sum total of these changes represent 'much ado about nothing'. Changes include that BNOC is no longer automatically awarded beneficial equity interests in all future licenses and that it has been made liable to the Petroleum Revenue Tax.[45] BNOC also surrendered about 25 per cent of its concessions to private companies because it concluded that it was overextended on exploration both financially and in terms of technical manpower.[46] At the same time, the government confirmed BNOC's continuing access to petroleum through options under participation agreements. In fact, it seems that 'the Tories are learning to love BNOC.'[47] The reasons for this 'love' are apparent, especially given the inadequate financial performance of other British state enterprises. Through BNOC, the government was sitting on a large pool of oil and – thanks to OPEC and its increased oil prices – the government found on its hands a profitable enterprise. During 1979, BNOC made a pre-tax profit of £75 million on sales of £3,245 million having sold nearly 1 million barrels of crude per day.[48]

In the beginning of 1980, BNOC's employment stood at 1,423, 85 per cent of whom were located in Scotland. There was every evidence that the oil multinationals were eager to enter into possible partnership discussions with BNOC in the seventh round of licenses in 1980. BNOC's relationships with the oil MNCs were close and cooperative. The sixth round, held in March 1979, saw BNOC participate in twenty-three of the twenty-six blocks bid for. 1979 was a milestone for BNOC in other aspects as well. It made its first entry into direct sales by selling fuel oil and LPG into the domestic market. (It has, of course, been a very large trader of crude oil.) Also, it concluded a technical service agreement for offshore exploration with the Malaysian national oil company, thus beginning the process of multinationalization. 1979 was a milestone for the government as well *vis-à-vis* its Common Market partners. Hitherto the government went to great pains in stating that it would not favor its domestic market at the expense of its EEC partners in case of a shortage. Yet in the summer of 1979, the Energy Secretary, David Howell, informed his EEC counterparts that the United Kingdom has the right to be protected from the shortages.[49] This represented a turn-around, especially from a government which pledged itself to reduce state intervention in the petroleum industry.

Probably the greatest change for BNOC at the end of the seventies proved to be the retirement of its founder chairman, Lord Kearton. Kearton, 68, wished to retire and there was no suggestion that his leaving had anything to do with the change in the government. He was replaced as chairman by R. E. Utiger, who was a BNOC director since the company's inception in 1976. Utiger, 54, was educated at Oxford and was the chairman of the (private) British Aluminium Company until November 1, 1979, when he took over BNOC's helm from Lord Kearton.

North Sea Oil and the British Economy

Numerous estimates have been made about the impact of North Sea oil on the British economy. The official version was given by the Treasury.[50] However, as of Spring 1977, one of the most comprehensive estimates available – making a

Table 2. North Sea oil contribution to the UK balance of payments (US $ billion at current prices)

	Proved and probable (P + P)	P + P plus possible
1976	1.3	1.3
1977	2.8	2.9
1980	10.6	10.9
1985	20.5	31.7

number of output and price assumptions – gave the results shown in Table 2.[51] In this case, the authors assumed the early 1977 crude oil price and an annual price increase of 5 per cent until 1985. Translating the balance of payments gains into potential GNP, in 1980, oil revenues would account for 3–4 per cent of GNP and in 1985 for 3–7 per cent of GNP. By any measure of means, this is a significant contribution.

Assuming oil at about $14.75 per barrel (unchanged through 1985) another study estimated that contributions to the balance of payments will be about $16.2 billion by 1985. The cumulative benefits are estimated at $90 billion while Britain's net overseas indebtedness during 1967–76 period was less than $11 billion.[52] According to this estimate, GNP would increase at a minimum an extra 1 per cent annually and, more likely, at an extra 2.25 per cent annually with the oil revenue. (By April 1980, the approximate price of North Sea crude was $30 per barrel and thus the above calculation should be adjusted accordingly.)

A more conservative estimate has the North Sea oil contributing 0.5 to 1.0 per cent incrementally to GNP growth in the 1980–85 period.[53] This study also projected that Britain will be self-sufficient in oil for only about a decade.

The flow of revenue will depend substantially on the production profile; a profile which had been changing even within a relatively short time (see Table 3). With time, the forecast has been increased for the short term while the longer-term forecast has become somewhat more cautious.

The further in the future the projections are made, the wider is, of course, the gap between the different forecasts (see Table 4).

In 1975 the net effect of North Sea oil on the UK balance of payments was £900 million and was estimated to rise to £2.1 billion in 1977 and about £5.0

Table 3. Forecast petroleum production[54] (millions of tons)

	1976	1977	1978	1980
Forecast made in July 1976	15–20	35–45	55–70	95–115
Forecast made in April 1977	–	40–45	60–70	90–110
Actual	12			

Table 4. UK North Sea: forecast oil production[55] (millions of tons)

	1980	1985
Government, April 1977	90–110	100–150
Robinson and Morgan, November 1976	102	135
Wood Mackenzie, February 1977	114	137

billion in 1980 – in constant 1976 prices.[56] Just as importantly, by 1985, North Sea oil will give the government about £4 billion a year in extra revenue.[57]

The implications for the oil companies – as contrasted to the British economy – of North Sea oil is that – because of generous capital allowances – the payback period for an oilfield is less than five years from the first production. The risk-adjusted discount rate is 17–21 per cent after tax, and using this criteria, twelve out of the fifteen fields discovered are profitable.[58]

UK Energy Consumption

The importance of North Sea oil to the UK economy can be further appreciated if the role of oil, from all sources, is examined. As late as 1970, coal provided more energy than did oil. In fact, the United Kingdom moved from a coal-based to an oil-based economy just at the time when the price of oil quadrupled. The displacement of coal by both oil and gas was very swift indeed (see Table 5).

All major industrialized countries but the United Kingdom are partially attempting to reduce their dependence on imported oil by increasing electricity generation from nuclear energy. Without fail, in every country the nuclear program is embroiled in environmental and safety controversies. It is an extra bonus for the UK public policy that North Sea oil and gas has saved the United Kingdom from this controversy, or at least did so for the foreseeable future.

Table 5. UK energy balance sheet (percentages based on million tons coal equivalent)

	1970	1975	1980 (estimates)[59]
Coal	46.7	35.2	36.0
Oil	44.5	45.0	36–41
Natural gas	5.3	16.1	17–22
Nuclear	2.8	3.1	5.0
Hydroelectricity	0.7	0.6	1.0

Appendix 1. Offshore Operations – Special Tax Provisions

(a) PRT will be levied on total revenues, less royalties and operating expenses.
(b) Oil companies will be allowed to write off total capital expenditure on a field, plus a 75 per cent uplift, plus any abortive exploration expenditure on that field before tax becomes payable.
(c) Oil companies will be allowed an oil allowance of 1 million tons of oil per year per field (20,000 barrels per day). This threshold exemption will apply up to a cumulative total of 10 million tons production from the field.
(d) Companies will not become liable for PRT if their pre-corporation tax return on their total capital expenditure falls below 30 per cent. In addition, there is a 'tapering provision' which in effect means that a 30 per cent return on capital before tax (about 15 per cent after tax) cannot be taxed at greater than 80 per cent. If a company does not make 30 per cent return on capital, it does not pay PRT. It would still pay corporation tax.

Points (b), (c), and (d) are the three major concessions which constitute the 'safety net' to protect the smaller fields, to guard against the possibility of the world oil price collapsing, or development costs getting out of control.

Appendix 2. UK Oil Marketers, 1975 (per cent)

Name	Approximate market share	UK refinery capacity	North Sea finds
Esso Shell	20–25	20–25	15–20
BP	15–20	20–25	15–20
Mobil Texaco	5–10	5–10	2–5
Total Gulf Fina	2–5	3–5	1–3
Conoco Amoco Phillips Chevron Burmah Murco Elf Occidental	2	0–4	1–4
All others	5	0	Approx. 25 (incl. nonoil interests)

Appendix 3. Estimated Offshore Oil Production Levels from the UK North Sea

Field	Operator	Estimated recoverable reserves (m.bbls.)	Production levels (thousand barrels per day) 1977	1980	1985
Argyll	Hamilton	70	30	24	–
Auk	Shell	50	40	13	–
Beryl	Mobil	400	70	80	65
Brent	Shell	2,000	50	400	385
Claymore	Occidental	410	40	140	70
Cormorant	Shell	160	–	40	36
Dunlin	Shell	400	–	80	100
Forties	BP	1,800	450	450	253
Heather	Unocal.	150	–	50	31
Montrose	Amoco	180	35	50	31
Murchison	Conoco	288	–	16	78
Ninian	Chevron	1,000	–	275	200
Piper	Occidental	800	150	220	130
Statfjord	Conoco	438	–	15	65
Thistle	BNOC	450	10	180	80
Total		8,596	875	2,033	1,524
Estimated production probable finds		5,010	–	248	1,225
Total production		13,606	875	2,281	2,749

Source: Section Two, *North Sea Service*, Wood Mackenzie, February 1977.

Bibliography

1. House of Lords, *Petroleum and Submarine Pipelines Bill*,
   2nd reading, Parliamentary Report No. 139, August 7, 1975.
   Parliamentary Report September 24, 1975.
   Parliamentary Report No. 155, October 29, 1975.
   Parliamentary Report No. 157, October 31, 1975.
2. House of Commons, *Petroleum and Submarine Pipelines Bill*,
   Parliamentary Report No. 193, November 5, 1975.
3. *Oil Taxation Bill*, November 14, 1974.
4. *Labour Party Manifesto*, October 1974.
5. *North Sea Oil and Gas*, First Report from the Committee of Public Accounts, February 14, 1973.
6. House of Lords, *Energy Bill*, As amended by Standing Committee J, pp. 1–27.
7. *Development of the Oil and Gas Resources of the United Kingdom*, Report to the Parliament by the Secretary of State for Energy, HMSO, 1977.

References

1. Roberts and Probert, 'The Effect of North Sea Gas on the Growth and Strategy of the British Gas Industry', *Revue de l'Energie*, November–December 1975, pp. 85–93.
2. Department of Energy, *Development of the Oil and Gas Resources of the U.K., 1976*, London, HMSO, April 1976. Latest estimate, used here, *Oil and Gas Journal*, January 1, 1977.
3. For an ideological background, see Peter Odell, *Oil the New Commanding Height*, Fabian Research Series, 251, London, December 1965. Odell advocated state direction but termed state ownership (he must have meant nationalization) 'both impracticable and unnecessary' (p. 17).
4. Adrian Hamilton, 'U.K. Policy and the North Sea', *Revue de l'Energie*, November–December 1975, pp. 144–152.
5. Lord Balogh, 'The North Sea Oil Blunder', *The Banker*, March 1974, p. 283.
6. Adrian Hamilton, 'U.K. Policy and the North Sea', *Revue de l'Energie*, November–December 1975, pp. 144–152.
7. Lord Balogh, 'Petroleum and Submarine Pipelines Bill', *Hansard*, House of Lords, July 23, 1975, pp. 334–335.
8. Report by E. Dell. See 'The North Sea Oil Blunder' by Lord Balogh, *The Banker*, March 1974, p. 281.
9. Adrian Hamilton, 'U.K. Policy and the North Sea', *Revue de l'Energie*, November–December 1975, p. 148.
10. The Rt. Hon. Patrick Jenkin, MP, 'The Conservative Policy for Off-shore Oil', *Petroleum Review*, September 1974, pp. 569–573.
11. Presentation to a meeting of financial analysts by Sir Frank McFadzean, Managing Director of Royal Dutch/Shell on November 22, 1973, p. 10. For further background see *Oil and Gas from the North Sea*, Shell Briefing Service, London, September 1974.
12. Sir Kenneth Younger's memo in Anthony Sampson, *The Seven Sisters*, Hodder and Stoughton, London, 1975, p. 120.
13. For a detailed discussion see Guy de Carmoy and Martin Flash, *North Sea Oil, Britain and Scotland*, INSEAD, Fontainebleau, 1977, mimeographed.
14. Eric Varley, Secretary of State for Energy, 'Petroleum and Submarine Pipelines Bill', *Hansard*, House of Commons, April 30, 1975, p. 483.
15. Adrian Hamilton, 'Tax Safeguards that Lift Some of the North Sea Clouds', *Financial Times*, February 26, 1975.

16. Adrian Hamilton, *Financial Times*, February 26, 1975.
17. *Petroleum and Submarine Pipelines Act 1975*, Chapter 74, Schedule 2, pp. 85–91.
18. *Petroleum and Submarine Pipelines Act 1975*, Part III, Section 23, pp. 21–23.
19. *Petroleum and Submarine Pipelines Act 1975*, Part III, Section 22, pp. 20–21.
20. *Petroleum and Submarine Pipelines Act 1975*, Part I, Section 2, pp. 3–4.
21. *Petroleum and Submarine Pipelines Act 1975*, Part I, Section 2, pp. 3–4.
22. White Paper, *U.K. Offshore Oil and Gas Policy*, CMND 5696, July 1974.
23. Taken from *Petroleum and Submarine Pipelines Act 1975*; summarized in Jonathan Story *et al.*, *North Sea Oil: Its Effects on Britain's Economic Future*, INSEAD, Fontainebleau, June 1975, pp. 40–41.
24. *International Herald Tribune*, June 17, 1977.
25. Ray Dafter, *Financial Times*, February 27, 1976.
26. Anthony Wedgwood Benn's speech to the British Manufacturers of Petroleum Equipment, London, February 13, 1976. *Petroleum Review*, March 1976.
27. Ray Dafter, *Financial Times*, May 28, 1976. For details see *Government Participation UK Style; Some Legal Aspects* and *Participation Agreements – BNOC*, International Bar Association, London, 1978.
28. Benn's speech, see reference 26.
29. *The Times* (of London), July 7, 1976, p. 21.
30. *Daily Telegraph*, July 8, 1976, p. 12.
31. 'U.K. Fifth Round', *Noroil*, February 1977, p. 19.
32. The Rt. Hon. Patrick Jenkin, MP, 'Petroleum and Submarine Pipelines Bill', *Hansard*, House of Commons, April 30, 1975, p. 503.
33. See *The Economist*, May 11, 1974, p. 65. Also for objections of MNCs to the BNOC concept and the tenuous state of participation as late as end 1975 see *The Economist*, November 1, 1975, pp. 68, 69.
34. *Financial Times*, April 10, 1975.
35. The Rt. Hon. Patrick Jenkin, MP, 'The Conservative Policy for Off-shore Oil', *Petroleum Review*, September 1974, pp. 569–573.
36. In late 1974, the Bank of England extended extensive guarantees and it purchased Burmah Oil's interest in BP.
37. For speculation that BNOC will follow the French Elf-Aquitaine model, see *Petroleum Economist*, July 1977, p. 280.
38. *The Economist*, March 6, 1976, p. 86.
39. *BP Statistical Review of the World Oil Industry, 1975*.
40. BNOC, *Report and Accounts 1976*, London, May 1977, p. 14.
41. Ray Dafter, 'Big Response to BNOC Staff Hunt', *Financial Times*, August 2, 1976.
42. Adrian Hamilton, 'BNOC Makes Its Presence Felt', *The Financial Times*, December 7, 1976, pp. vi–20.
43. BNOC, *Report and Accounts 1976*, London, May 1977, p. 15.
44. Private communication to author, June 1977.
45. The PRT has been revised in August 1978. See C. Robinson and C. Rowland, 'Marginal Effect of PRT Changes', *Petroleum Economist*, December 1978.
46. *Financial Times* (London), October 2, 1979, p. 20.
47. *The Economist* (London), June 9, 1979, p. 111.
48. BNOC, *Report and Accounts*, 1979, pp. 4–5.
49. Private communication to author, Fall 1979.
50. 'North Sea and the Balance of Payments', *Economic Progress Report*, July 1976.
51. Colin Robinson and Jon Morgan, *Effects of North Sea Oil on the U.K. Balance of Payments*, Guest Paper No. 5, Trade Policy Research Center, November 1976.
52. Private communication to author, 1976. We assumed a sterling exchange rate of $1.80 through 1985; the figures were given in sterling.
53. British Petroleum, *Some Effects of North Sea Oil on the U.K. Economy*, Policy Review Unit, Occasional Paper, October 1976.

54. Banque de Neuflize, *L'Economie Britannique et le Pétrole*, Schlumberger, Mallet, Paris, July 1977, p. 3.
55. *Petroleum Economist*, June 1977, pp. 217–219.
56. *International Herald Tribune*, August 11, 1977.
57. *The Challenge of North Sea Oil* (White Paper), HMSO, London, March 1978.
58. P. Richards, R. Goodfellow, and A. Contesse, *Government Policy in the North Sea and Its Implications for Financing*, London Business School, June 1, 1975. Mimeographed, two volumes, and an update by Colin Robinson, 'North Sea Investment and Profitability', *DNC Oil Now*, Den norske Credit-bank, Oslo, November 1979, pp. 1–22.
59. 1980 estimate from Banque de Neuflize, *L'Economie Britannique et le Pétrole*, Schlumberger, Mallet, Paris, July 1977, p. 2.

CHAPTER 8

STATOIL: Norwegian National Oil Company

STATOIL, the Norwegian national oil company, is the second youngest national oil company in Europe, and is small when compared with the others. Indeed as yet it is not large in Norway. But as the principal operator in the big oilfields of a country whose needs for the oil are small, STATOIL has the potential to become an important enterprise both nationally and internationally. Unlike the other Western European countries, Norway has no large internal oil market and has plentiful indigenous supplies. Present discoveries alone, made on a geographical area covering less than 7 per cent of the Norwegian Continental Shelf, are enough to supply the internal market at existing consumption rates for the next 160 years. Already a wealthy country with deep concern for its style of development, Norway has small need of oil revenues to bolster its economy. This sets it apart from its North Sea partner, the United Kingdom, where the oil revenues are needed to finance debts and domestic requirements as rapidly as technology will allow. The Norwegians want to develop their new-found resources slowly in order to disturb the Norwegian way of life as little as possible.

The case of STATOIL is not only of interest for its illustration of the problems of building a new organization capable of handling itself in the international oil industry; it is also a study of conflicts of interest. Norway's allies in Western Europe see Norwegian oil as a possible major contributor to the reduction of their dependence on the Middle East oil sources and would like to see Norway's resources fully developed. The Norwegians would like to control development. The large oil and gas discoveries made so far reinforce the desire of the government to develop the Norwegian oil industry steadily, but the oil industry is now geared for a substantial work load. Environmental considerations of the impact on the economy in general and on the agriculture and fishing industries in particular also have an important bearing on the pace and extent of development. Finally there is a potential conflict between the government which wants to control STATOIL, and STATOIL which wants to be free from political control and which sees in the immense oil revenues it may have at its disposal the means by which it will achieve this. The question of who is going to control whom is paramount.

Norway

Norway is not a large country *per se*, except relative to the population of only just over 4 million. Its land area is extremely long, 1,100 miles, and is mountainous, with a small agricultural area (0.3 per cent). Much of the population is concentrated in the South around the capital Oslo (470,000) and the cities of Bergen (234,000), Trondheim (128,000), and Stavenger (95,000) but small settlements exist all along the coastline. The topography enables 55 per cent of the energy to be supplied by hydroelectricity, and with 5 per cent from other sources, only 40 per cent is supplied by oil. The economy embraces a large amount of primary industry (fishing, timber, agriculture) which employs 10 per cent of the active population. Shipbuilding, cement, and aluminum are leading industries, exports play a very important role, and the Norwegians have become the wealthiest nation in Europe.

Norway is the only NATO country, other than Turkey, to have a land border with the USSR (see Appendix 1). Norway is not, however, a member of the European Economic Community (EEC) as the result of a referendum taken in 1972, but is a member of both the European Free Trade Area and the Nordic Council. Although the precise composition of the government has changed from time to time, Norway belongs to the group of social democratic governments of Northern Europe, and a program of social welfare is complemented by a direct taxation structure and government intervention in a number of industrial fields, notably power generation and shipbuilding. A traditionally outward looking country with a long seafaring experience, Norway nonetheless stands alone. There is no immigration and the rest of Europe is treated as 'the Continent', but many Norwegians take their further education overseas and work there for a period. Norwegian society is one of well-established and strongly held values, and the discovery of oil on the Continental Shelf in large quantities appeared not so much as a long sought after bonanza, but a change in the environment which could be as much a threat as a potential benefit.

Early Developments

The North Sea had, for a long time, been considered an area of potential hydrocarbons, and interest was further stimulated by the discovery of gas in the Netherlands at Groningen[1] in 1958. The first approach to the Norwegian government to explore the Continental Shelf was made in 1962, but no immediate reply was given as it studied the problem on division of sea bed resources. In 1963, after declaring sovereignty over the sea bed, various licenses for seismic exploration were issued, and a series of agreements apportioned the areas of the North Sea to the adjoining countries.

Legislation defining exploration and production rules was passed in 1965, and production licenses were allocated on seventy-eight blocks to nine companies.[2] Administration of the government's interests was initially handled by the Mining Division of the Ministry of Industry but in 1966 the volume of work gave cause to create a separate Petroleum Division.[3] Policy development by this body led to the stipulation that state participation was an obligatory provision for the next licensing of exploration areas, which were awarded in 1969–71 over fourteen blocks to six groups, all already operating in the Norwegian offshore area.

The Arab–Israeli war of 1967 quickened interest in the North Sea and in 1969 the first significant find was made at Ekofisk by the Phillips group[4] (see Appendix 2). This served to underline the fact that much of the work of the Petroleum Office needed reorganization in recognition of its technological and commercial character, and accordingly a committee was appointed to make recommendations. It was chaired by K. E. Knudsen, who subsequently became President of Saga Petroleum, a wholly private company. The committee reported in 1971, the year when the major gas strike at Frigg[5] was made, and as a result the government made three proposals, which were accepted by the Storting (the Parliament) in June 1972. These were to establish a Norwegian Petroleum Directorate, to form a separate Petroleum and Mining Department at the Ministry of Industry, and to establish a national oil company, STATOIL, both as a holding and operating company.[6] (The committee had recommended only a holding company.)

The Establishment of STATOIL

In September 1972 STATOIL was formally founded with the following objectives:[7]

> To carry out, by itself or in participation or in cooperation with other companies, exploration and production, transportation, refining and marketing of petroleum and products derived therefrom, and other activities reasonably related thereto.

Subject to the normal provisions of the Norwegian Companies Act the company nonetheless was expected to fulfill objectives larger than those of private enterprise, responding to the political and social aims of the government as well as to its economic ones. These aims were assuring supply security, ensuring optimal development, acquiring expertise and technology, providing better control of the private operators, fostering competitive Norwegian goods and services, and earning a reasonable return on the government investment. In effect, STATOIL was charged with the caretaking of the state's petroleum interests, but its freedom to interpret its brief was curtailed by its Articles of Association, which read:[8]

> The Board shall submit to the General Meeting (the Minister of Industry), ordinary or extraordinary, all matters which are presumed to involve significant political questions or questions of principle and/or which may have important effects on the nation or its economy. Such matters shall be deemed to include, *inter alia*:
> (a) Plans for the next following year with economic surveys, including plans to cooperate with other companies.

(b) Essential changes of such plans as mentioned in (a) above.
(c) Plans for future activities, including participation in activities of major importance in other companies or joint ventures in which the company participates or plans to participate.
(d) Matters which seem to necessitate additional appropriation of government funds.
(e) Plans for establishing new types of activities and localization of important elements of the company's operations.
(f) Plans to participate in exploration of petroleum resources in or outside Norway, including the exercise of government participation option rights.
(g) Semi-annual reports on the company's operations, including operations of subsidiaries and joint ventures with other companies of importance.

Notwithstanding the restrictions on its freedom and the ambiguities of its role, STATOIL's initial preoccupation was with establishing itself. The first staff, the director and a secretary, started work on January 1, 1973, in Stavanger where the headquarters had been placed to be close to the oil activity. Stavanger was also the seat of the new Petroleum Directorate which arrived the same year with the intention of supervising and controlling all activities on the Norwegian Continental Shelf, enforcing the regulations and negotiating license awards for drilling and production pipelines.

In May of 1973, STATOIL was assigned two options, a 5 per cent interest in Frigg and 40 per cent in Heimdal, and four months later a 50 per cent option in Statfjord (see Appendix 3). They were a fortunate start, for that year East Frigg was declared a major gas discovery, and in 1974 Heimdal was provisionally similarly classified and Statfjord doubled Norway's proven reserves overnight to 8 billion barrels.[9] Thus within two years of its founding STATOIL was already the part owner of very substantial reserves of gas and oil. During this time the staff of the company had grown rapidly and it was to continue to do so (see Appendix 4).

The means by which STATOIL took up its participation in the oil concessions evolved through the period from 1973 to 1975. Prior to 1973, various forms of state participation had been used. One of these was a 'net profit' formula, giving to the government a certain percentage of the licensee's net profits and rights to information but not real additional influence other than that exercised by government in its administrative capacity.[10] Another type of agreement was the 'carried interest' agreement whereby the government had the option to become a partner on equal terms with the other licensee(s) if a commercial find was made. At this point the government became liable both for its share of the exploration costs (including in some cases costs prior to the option being exercised) and for development costs. As from 1973, only the 'carried interest' type of agreement has been used, with some further modifications to the formula described above. Under these agreements, STATOIL never pays the exploration costs. If a discovery is accepted by STATOIL as being commercial, it will from then on pay its share of the development costs. As from 1975, STATOIL's minimum interest is 50 per cent in any concession and once a production profile is determined, participation can be increased up to a maximum share which will be determined in accordance with an agreed table based on the size of the production.[11] If a major oilfield is discovered the final STATOIL share will be in the area of 60 to 75 per cent. As a majority participant, STATOIL will have the decision-making powers subject to certain minority protection clauses. In many cases STATOIL has an option to assume the role of operator in the concession area and in four more recent cases STATOIL is the operator from the outset, with technical assistance being rendered by a foreign oil company.

The quick growth of its staff and sudden acquisition of large reserves, however, had provoked concern in the Storting over the growing influence of the company, and moves were made to restrain it.

STATOIL and the Government

STATOIL's managing director, Arve Johnsen, was a forceful personality with ambitions to create a fully integrated international oil company. Born in 1934, he had been educated in Bergen and the United States in economics, business administration, and law, and had worked with the Norwegian Export Council and Norsk Hydro before joining STATOIL. He had campaigned actively for Norway joining the Common Market and considered his experience as a politician the most formative for his position in STATOIL, enabling him to understand how decisions are made and making him realize how, in his words, 'power is basic – it is basic at all levels.' There was some feeling that he had the intention of running STATOIL both as a business and a political instrument.

The government, for its part, however, was all too aware of this, and was concerned from the start to limit STATOIL's power. The legislators had in mind the experience of Italy with Enrico Mattei and the struggle to get ENI controlled by the government and not vice versa. Arve Johnsen was felt to have a style (and ambition) that was altogether too 'buccaneering'[12] and to curb his power over the Ministry of Industry a new 'heavyweight' Minister was placed there in late 1975. Also STATOIL's right to borrow money (with government guarantees) was withdrawn, enabling the government to scrutinize all capital requirements.[13] The pressure for curbing STATOIL came from an odd coalition of left and right wing forces. The former, notably the Socialist Party with a controlling minority in the Labour government, wanted STATOIL to be more 'accountable'. The latter, particularly the Civil Service, the Ministry of Finance, and the Ministry of Industry, did not want another center of power to deal with.

Of course, the original Articles of Association (see the section 'The Establishment of STATOIL') had from the start subordinated STATOIL's commercial objectives to national ones, but STATOIL was not a passive organization and had naturally become an important contributor of facts, evaluations, and viewpoints regarding activity on the Continental Shelf to the Ministry of Industry. As the state's own business organization it entered into dialogue with the government on any important matter. In theory STATOIL was to act within the guidelines established as a result of the political debate concerning the activities and future direction, but STATOIL had ambitions of its own and thus sought to influence the very political debate which was its ultimate control. What was of central interest to STATOIL was of course the extent to which it would be allowed to develop into a fully integrated oil company, and the degree to which it would be master of its own destiny.

The Vertical Integration of STATOIL

The acquisition of petroleum technology for exploration and production was to be undertaken by STATOIL, and to this end it participated directly in the analysis and development of the Statfjord, Frigg, and Heimdal fields (see Appendix 5), as well as other exploratory offshore ventures, including one in conjunction with Conoco on the Dutch Shelf. In transportation STATOIL had taken a 50 per cent share of the gas and oil pipeline companies going to Germany and Britain respectively from the Ekofisk field, and its 5 per cent interest in the Frigg field gave it a similar share in the gas pipeline to the United Kingdom which started its operation in 1977. It was engaged in feasibility studies for an oil pipeline from Statfjord to Norway (technically difficult due to the deep Norwegian Trench off the coast) and a gas line from Norway to the Continent. Investment had also been made in a company to process seismic data and build up a data bank.

But the real interest for STATOIL was in forward integration, so that it could be fully integrated as foreseen in the Articles of Association from the outset. Initial integration in refining and marketing was limited, and the intention was that refining capacity should be developed in order to allow the company to market royalty crude in the form of refined products. From 1974 STATOIL had been receiving royalty crude from Ekofisk in order to gain marketing experience, and had been constrained to sell most of it in Norway for refining. This had prevented it getting the best prices possible because North Sea crude was a premium crude and Norway was not a premium market. STATOIL had also been hoping that it would be able to have a relatively free hand in marketing products other than crude, but the belief that this would happen took a jolt in 1976 when the Storting approved the government's plans for a new refining and marketing company which was the result of the state buying the Norwegian operations of BP.[14] In the new company STATOIL would have only a 15 per cent interest, 71 per cent would be held by the state, the Norwegian Co-operative Wholesale Society would hold 2 per cent, the Saga Petroleum Company would hold 5 per cent, and 7 per cent would be held by Norsk Hydro, itself 51 per cent owned by the state. The state holding in Norsk Hydro was the result of reclaiming assets owned by the Germans at the end of World War II. Norsk Hydro had a significant proportion of foreign shareholders (it was quoted on the French stock exchange) and therefore was not considered to be a suitable vehicle for state intervention in the oil industry. The new company was called Norsk Olje A/S (NOROL), and private oil industry circles were apprehensive that it might eventually take over all government-related business. The vertical integration efforts of STATOIL received a major boost when – as a result of a Storting resolution – STATOIL purchased for 200 million kroners from the government new share capital in NOROL. Thus as of 1980, STATOIL owned 73.6 per cent of NOROL.

NOROL's part-owned refinery had a capacity of 80,000 barrels per day and was one of only three in Norway (the others belonged to Shell, 70,000 barrels per day, and Esso, 110,000 barrels per day). The refinery was originally a joint venture by Norsk Hydro and BP, but the latter was bought out. The new shareholding of the refinery was NOROL 40 per cent, STATOIL and Norsk Hydro 30 per cent each. There was some opinion that the purchase of the BP network was not a good one. BP (Norway), with much the same market share (25 per cent) as Esso, employed 80 per cent more people in Norway and the presence of 50 per cent Norwegian interests in BP (Norway) appeared to a few observers to have made the price paid unreasonably high.

The precise brief for the new company was not clear-cut but it reflected the desire for state participation in the marketing of refined products and its chairman expected to develop NOROL's activities in Sweden and Denmark, where some division of task might be needed as Norsk Hydro was already established there. The government was, in fact, undecided on its royalty policy as to whether it could gain a greater benefit for the state by receiving cash or oil as royalty on oil production. By marketing crude or products directly, the government could provide a check on STATOIL's activity, and through its shareholding in NOROL the government had a direct outlet for any crude it might receive. Initially STATOIL had been allowed to supply 70 per cent of the refinery's requirements, equivalent to its own share of the equity plus that of NOROL, but this was not considered a permanent arrangement. The creation of NOROL, which would supply 24 per cent of Norway's oil products market and have 30 per cent of its service stations (again the result of buying out BP), would constrain STATOIL to market its products elsewhere outside Scandinavia. This would conform with the Articles of Association but would also require that the government allow STATOIL to devote its large cash flows to foreign investments. In 1977 NOROL's organization was still absorbing the results of the various changes of ownership and the profitability and efficiency of the new grouping against the competition of Shell and Esso was still to be demonstrated, but it was at least clear that one effect of the various cross-patterns of ownership and marketing arrangements was to contain the freedom of STATOIL's action.

Integration into the field of petrochemicals was also similarly constrained, although the decision not tò make STATOIL operator of the petrochemical complex was taken (1973) before NOROL was created.[15] This was not least because Norsk Hydro successfully convinced the government that its activity in this field should not be jeopardized. In consequence STATOIL only had shares in two petrochemical ventures, in neither of which it was the operator. An ethylene plant, I/S Noretyl, was to use natural gas liquids (NGL) from the Phillips group,

Table 1. Petrochemical plant ownership (per cent)

	I/S Noretyl	I/S Norpolefin
Norsk Hydro	51 (operator)	33.3
STATOIL	33	33.3
Saga Petroleum	16	33.3 (operator)

Amoco, and Norsk Shell in connection with government agreements in 1973 and in 1976 connected with the Ekofisk field. A polyethylene and propylene plant, I/S Norpolefin, was to use feedstock from I/S Noretyl. The shares were as shown in Table 1 and both plants became operational in 1979. STATOIL thus found itself able to integrate upstream fully, but downstream economic rationale had given way to the political need to restrain the power of the company, at least in the immediate future.

The Oil Potential and Strategic Considerations

By 1975 the government was already convinced of the need to control the pace of development by restricting the acreage for exploration, and allocations of blocks were less in the third concession round than in previous years (forty-seven applications for twenty-six blocks; eight awarded). But it was helped to this conclusion by the large reserves that had already been found. Estimates of the reserves varied so much that forward projections of production were subject to caution. For example the estimate of the Petroleum Directorate of the reserves on the purely Norwegian side of the Statfjord field were 295 million tons; the licensees' estimate was 455 million tons. By 1976 the Petroleum Directorate estimates of available production were as shown in Table 2.

These estimates were downward revisions on previous estimates due to technical difficulties and insistence on economical production rhythms and were well below the official ceiling of 90 million tons which was reckoned to be a rate in harmony with the Norwegian economy. Nearly half the total was to be supplied by Statfjord alone. Total recoverable reserves were calculated as 1,030 million tons of oil and at the originally envisaged production level (1.8 million barrels per day) this was equivalent to ten times Norway's domestic requirements. The government was thus well placed to negotiate optimal agreements with the private

Table 2. Oil production[16]

1980	62 million tons oil equivalent
1985	69
1990	55

oil companies and to control the speed of development. However, by 1976 pressure had built up for further extensions of exploration activity and the government was obliged to face a new problem.

Up to 1976 only 7 per cent of the Norwegian Continental Shelf had been explored, and all of this acreage was in the area below the 62nd parallel (see Appendix 1). About 80 per cent of Norway's Continental Shelf lay to the north of the 62nd parallel, and observers agreed that it represented a huge hydrocarbon potential. However, the farther north the location of an exploration area, the more sensitive it would become politically. The USSR had at Murmansk the largest naval base in the world and its only Western ice-free port. A forest of drilling rigs belonging to members of the North Atlantic pact (NATO) would not have been welcome on its only access route to the Atlantic. In addition, there was no agreement between Norway and the USSR on the boundary between their continental shelves in the Barents Sea, although the Norwegian island of Spitsbergen was being jointly exploited by Norwegians and Russians for its coal deposits. Western pressure on Norway for the development of this area would probably have been exchanged for support of the Norwegian position in any negotiations with the Russians.

In 1976, the immediate pressure was coming from Norway's own oil industry. Having geared up for the early boom, the industry was not in a mood to contract in the face of obviously substantial reserves, and both unions and management in all oil-related activities were pressing the government at least to maintain the pace of previous years by allowing exploration north of the 62nd parallel to commence. By 1976 the government was sufficiently convinced to prepare plans for both STATOIL and the private companies to work farther north, starting in the Tromsø area of Northern Norway. Although regional development reasons played their part in this decision, technological progress and the slack in the industry were other factors. But the more northerly communities were those whose economic and social infrastructure would be most dramatically affected by oil activity and the farther north the location, the greater the weather exposure and the greater the danger from drifting icebergs. Consequently the Storting postponed any start while environmental, strategic, and safety issues were further considered. The BRAVO accident and the collapse of an Ekofisk platform in early 1980 further postponed exploration north of the 62nd parallel.

Fiscal Control

The creation of STATOIL had given the government some active control of oil development. It was buttressed by both administrative and fiscal measures covering all the companies – STATOIL included. Administrative control was kept by the system of licenses for exploration, production, and construction, and to these were attached fees, royalties, and income taxes. Application fees were fixed at NKr15,000 and area fees slid from an initial NKr1,800 per square kilometer to NKr15,000 for the tenth year. The area fee will thereafter be NKr30,000 per square kilometer for the remainder of the license period. Royalties were charged on production, again on a sliding scale, from 8 to 16 per cent for production from 40,000 to 350,000 barrels per day. The royalty of gas is fixed at 12.5 per cent on the value of the production. To royalties was added corporation tax (which from 1975 onwards was 50.8 per cent) and then a special tax on the remaining income.[17] The net 'take' of the government was in the end 57–66 per cent for an average size field; conditions were less favorable for small fields. Companies were, however, able to make a number of deductions before they paid tax, including an investment allowance, the writing off of production and transportation systems and exploration costs, and a ring fence provision allowing the writing off of losses from onshore activities.[18] Finally net earnings on which taxes were to be based were to be calculated using a 'norm' price fixed every three months by the Petroleum Price Council. All these provisions were considered as striking a reasonable balance between state benefit and private profit, but the operator of the only field producing in 1976, Phillips at Ekofisk, was from the first dissatisfied with the norm prices, considering them 30 c/bbl too high[19] (see Appendixes 6 and 7 for details).

STATOIL's Present Status and Prospects

It was against the background of the administrative and fiscal controls of the government that STATOIL developed. By the end of 1976 Norway's policy of 'hasten slowly' seemed to contain much to be commended when viewed against the precipitate rush of the British to exploit their part of the North Sea, although their reasons for hurrying were much more compelling than those of Norway. The government's efforts to foster all aspects of petroleum technology in Norway by 1976 were largely paying off with expertise developing in many fields. The creation of STATOIL fitted well with the government's ideas of state control of an important new resource, and any discouragement this gave to private interests was not regretted in the context of 'hasten slowly'. That STATOIL was envisaged as something more than a pure oil company responding only to commercial considerations was probably the reason why Norsk Hydro, with its private interests, was not the chosen state vehicle. Although the government had carried out its policies slowly and had attempted to limit STATOIL's power to become a state within a state, by the end of the seventies strains were emerging.

On the general level there were many pressures to allow exploration above the 62nd parallel, most importantly from the unions and industry who foresaw a slow-down in total North Sea activity due to cost inflation and limited opportunity unless this area was opened to them. Western interests, too, were concerned that Norway should not leave its potential unexplored. From STATOIL, there was pressure to go faster. The team assembled around Johnsen was still growing rapidly and was impressively well educated (see Appendix 4) and very competent, with much previous oil industry experience. It wanted to get on with the job of expanding STATOIL and so was pressuring for faster development of its fields, and for Johnsen there was the very real interest that as soon as STATOIL could generate enough cash, principally from Statfjord, he would be freer of government control.

Statfjord in fact was the dominant influence on the future of the company. It was clear that development of the field was already running one year later than originally planned. Nonetheless the first platform was put on station in 1978 and phase 2 plans for production of Statfjord (two more platforms giving a total production capability of 900,000 bbl/d) were going ahead,[20] despite the fact that no final solution to the transportation system was expected until 1980. Total investment for production facilities was estimated at NKr25 billion,[21] of which STATOIL would have to provide NKr12.5 billion.

The 1977 oil price was $13 per barrel c.i.f. and it rose to $30 by 1980. Operating costs in the North Sea, although a closely guarded secret, were known to be high. Operating costs at Ekofisk were slightly below $4 per barrel and Statfjord costs were not expected to be higher. Capital costs and the transport system envisaged for Statfjord were, however, certainly expected to be higher than Ekofisk and hence the operating cost per barrel, according to a private communication estimate and including fees and interest, was expected to be about $5.50 per barrel. Over a twenty-year period at an average production of 500,000 bbl/day this would produce a total income of NKr153 billion, with a present value (at 10 per cent discount) of NKr65 billion. STATOIL's own estimate of the expected return on the investment in Statfjord was 20 per cent but details of the structure of the investment, loans or capital, were unclear, and the existing estimate of the price of alternatives (nonoil) fuels gave a large leeway of cost alterations.

In the meantime, STATOIL was going to the government for money, to finance its investments, which between the plans put forward in 1975 and 1976 had jumped by 50 per cent, almost entirely due to cost increases for the development of Statfjord. (See Table 3 for STATOIL's investments.)

The bulk of these investments was to be financed by loans with some increases in share capital. For example, of the capital requirements of NKr2,783 million for

Table 3. Distribution of STATOIL's investments (millions of Norwegian kroner)[22]

	1976	1977	1978	1979	1980	Sum
Joint ventures	2,158	2,162	3,967	3,835	3,300	15,422
Limited liability companies	162	26	3	–	–	191
Directly administered by STATOIL	97	108	140	52	43	440
Sum[a]	2,417	2,296	4,110	3,887	3,343	16,053
Investments in Statfjord	984	1,693	3,476	3,250	3,056	12,459
Statfjord investments as a percentage of the total investments	40.7	73.7	84.6	83.6	91.4	77.6

[a]For detailed breakdown see Appendix 8.

1977, NKr2,000 million was to come from credits and NKr500 million from a share capital increase (to a total of NKr2,015.5 million). The directors of STATOIL had asked for the share capital to be increased by this amount each year until 1980 to give the company the necessary solidity and, one might add, the desirable independence. But the Minister of Industry reported to the Storting on October 15, 1976 that 'one has not as yet decided on the future capital structure of the company'[23] and consequently no immediate shift of power to STATOIL could be foreseen while the government kept its hands on the purse strings.

Profits from STATOIL's investments were not expected to be generated until 1985, although project operations would make a positive contribution to financing from 1980 (see Appendix 9); but it was not clear, however, how the government would exercise budgetary and planning control. Although the question of whether a key should be adopted for the division of STATOIL's earnings between the company and the government had been widely discussed, the subject would remain of little practical interest until the middle eighties. To include in the company articles a provision relating to the division of STATOIL's gross earnings between the company and the government as shareholder was incompatible with the rules of the Companies Act.[24] The Act did not permit a shareholder in a joint stock company to secure for himself a direct share of the company's earnings. Should it later be found necessary to have an article providing that part of STATOIL's earnings should not pass through the company but go direct to the government, a special bill would have to be enacted.

By the eighties STATOIL might have acquired its own momentum. In the appointment of board members and key managers the government was motivated by consideration of ability and the curbing of Johnsen's real or perceived power, and the Board contained an ex-Minister of Finance, two members of the Storting, and a municipal administrator among its seven members (the other two were elected by and among the employees). Also the government had made sure that both the Petroleum Directorate and the Department of Petroleum and Mining in the Ministry of Industry had kept pace with the growth of STATOIL and oil activity in general. By the end of 1976 the staff of the former was over a hundred and that of the latter was nearly forty, of whom twelve were university educated economists and fifteen were lawyers (see Appendixes 10 and 11). But as mentioned above the overall aspect of the STATOIL staff was ability, experience, and international outlook with many having shared educational backgrounds. The accumulation of talent at Stavanger, at the headquarters of STATOIL, was viewed by many Norwegians as a somewhat necessary evil – necessary, because Norway would have to market 90 per cent of its oil and gas abroad and competence would ensure that the country would sell it at optimum terms, and an evil because the combination of money and talent would – without a doubt – create a very powerful organization indeed (see Appendix 12).

The Economy, the Impact of Oil and Gas, and STATOIL

Norway's economy had benefited and would continue to benefit enormously from

Table 4. Impact of oil and gas on Norway's economy

	Millions of NKr	Millions of $
Exploration expenses	2,800–3,000	500–550
Average well cost	20	3.6
Investment in oil, up to 1976 (in 1975 prices)	40,000	7,300
Estimated investments 1976–83	80,000	14,500
Likely annual income flows[a]	10,000–15,000	1,800–2,700

[a] Income flows actually exceeded the above estimates notwithstanding the lower production levels because of the large increases in crude oil prices which, in early 1980, were about $30 per barrel.

the oil boom. Figures provided by the Den Norske Creditbank showed this impact dramatically (see Table 4). As a basis of comparison the entire income of the government, including some oil and gas revenue, was NKr34 billion. The government would only receive the tax-take proportion of the annual income flows plus any income from STATOIL and NOROL, but this could amount to more than 12 per cent of its total income.

The oil flows envisaged at the end of 1979, the value of which would depend on both estimated oil production (see Appendix 13) and the development of prices, would last at least thirty-five years. The gas supplies, essentially only from the Frigg and Ekofisk fields because Heimdal was still under consideration as a commercial development, were likely to last at least fifty years. By 1985, it was expected that the overall gas consumption of West Germany, France, and Benelux would amount to 171 billion cubic meters of which Norwegian exports would account for more than 17 per cent, and in the United Kingdom, Norway would be supplying 15 per cent of consumption. More than three-fifths (400 billion cubic meters) of Norway's estimated gas reserves (see Appendix 14) had been committed for sale.

By the end of the seventies Norway had experienced some economic difficulties, however. In the process of developing oil, Norway had to borrow heavily internationally and, simultaneously, curb domestic spending. Mortgaging future oil revenues, Norway embarked on a policy of high wages, runaway consumption, a deteriorating balance of payments, and high debts.[25] One result of Norway's large indebtedness was that in 1978 STATOIL was encouraged by the government to borrow directly because of the preference of banks to lend to a commercial concern rather than to a government with increasing indebtedness.[26] Chances are, of course, that STATOIL itself will not be profitable before 1982 and will not be able to clear its own debts before 1985 or even after. In the course of 1979, Norway's economic situation improved due to the devaluation of the kroner, a wage and price freeze, and (further) measures to subsidize exports.

Another variable on the Norwegian scene was that oil development had proceeded more slowly and proved to be more costly than was anticipated. By the late seventies policy still was to reach production goals that do not exceed 90 million tons of oil a year, even though the realistic target was about 60 million tons.

Therefore, the government adopted a flexible licensing schedule;[27] for instance, if the fourth round (1978–79) did not meet the government's expectations, it would begin to think about allocating blocks under the fifth round. Terms for the fourth round were similar to the third round (1974–75) with STATOIL obtaining a 50 per cent share in each concession (which can rise to between 70–80 per cent depending on the size of the discovery). Additionally, the fourth round licensees were expected to adhere to the government's 'buy Norwegian' policy to help to alleviate the country's poor balance of payments situation. The oil policy White Paper[28] published in early 1980 confirms that 'Norwegianization' of the oil industry will continue with STATOIL continuing to play a pivotal role, Norsk Hydro and Saga Petroleum important roles, and with the foreign companies acting as contractors with payments of fees rather than crude oil. Whether the interest of the foreign companies may be maintained under these conditions is a matter of conjecture.

Assessment as of 1980

The Norwegian government oil policy had an impressive record of clarity, consistency, and an ability to learn from mistakes. STATOIL had been established both thoughtfully and capably. Care had been taken not only to clearly define its field of action but also to ensure that it was provided with competent personnel. The government bodies were also well briefed and staffed and the policy lines were clearly laid out. The government had come down in favor of a certain concentration of Norwegian petroleum industry: STATOIL, the large Norsk Hydro, and Saga (a private consortium of some ninety companies and 50,000 shareholders). It has also stuck to its 'hasten slowly' policy which was intended to keep oil activities in harmony with Norway's existing economic development and social organization. The government had not allowed itself to be hurried by others and its declared, and to date, enacted policy had been one of strong central control. Much as this went against the concept of free enterprise, one could not help admiring the results achieved in Norway.

STATOIL, as seen above, still claimed in public that it had no separate voice and that it was an instrument of the government, but privately its voice was heard increasingly on the side of industrialists pushing for a faster rate of exploitation. Unlike some national oil companies in the past, STATOIL had started life with crude supplies. The question now was whether it would be able to follow the ENI example: integrating vertically forward, entering into large-scale international agreements and operations, but without taking on noncommercial social responsibilities. The Articles of Association tended to indicate this would be difficult without the government's sanction.

For the future, the crucial question remained as to what STATOIL was going to do with its excess crude and its share of profits from 1982 onward, especially if and when future production came on stream from possible new discoveries both above and below the 62nd parallel. Under the established Norwegian oil policy,

STATOIL was automatically a majority participant in all new concessions, and would thus have access to crude oil according to its participation interest.

As of the late seventies, a four-way conflict seemed to be developing between STATOIL, the government, and the political parties outside the government and private industry. STATOIL wanted to become a fully integrated multinational company operating like a 'major'. (Because of the large Norwegian shipping interests, STATOIL would not become an owner but a charterer of tankers.) The Labour government and the civil service were anxious to keep close control on the company, and the non-labour parties and private industry were ready to slow down STATOIL's rate of growth and power. The future, of course, was a matter of conjecture, but present odds remained with the Norwegian government keeping STATOIL firmly under its commercial and political control.

In 1980, STATOIL reached its eighth birthday. It seemed that some of the early frictions have subsided and that STATOIL and the government worked together obtaining each other's approval. STATOIL managed to maintain its freedom of managerial action – given the legal and regulatory policies laid down in the cumulative licensing rounds. Arve Johnsen was still president and very much in control of STATOIL.

Appendix 1. Norway – Geographical Position

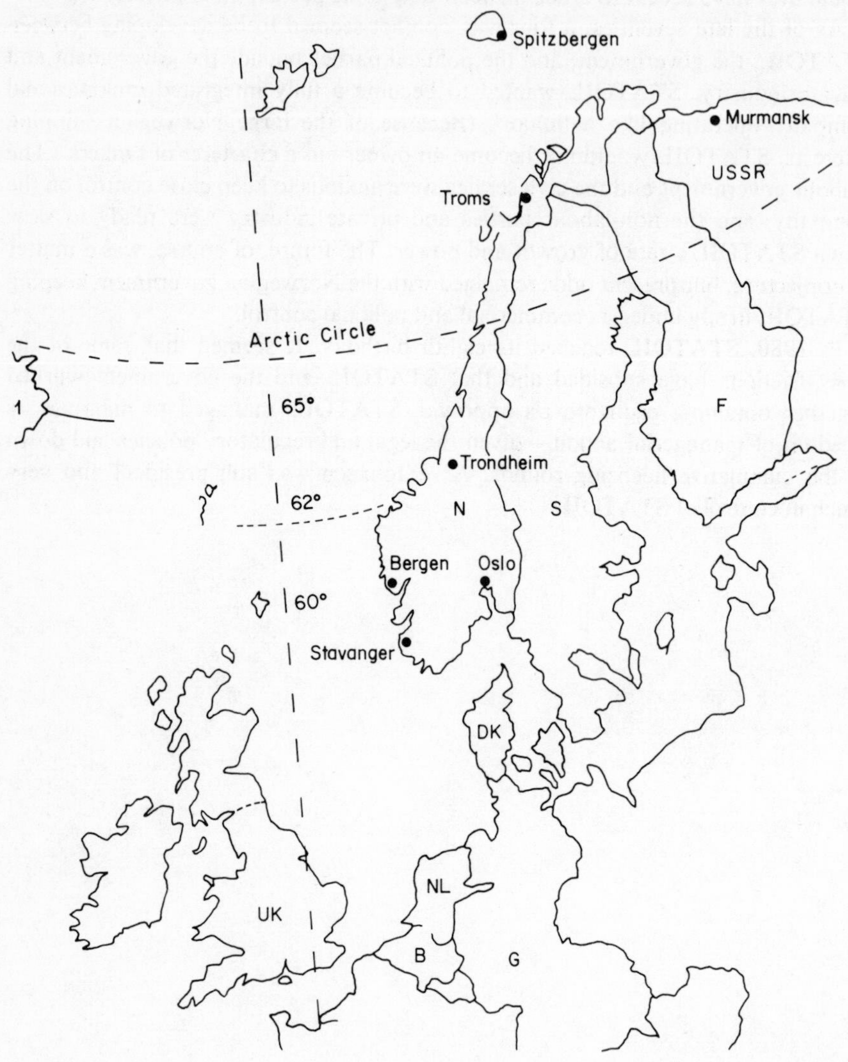

Appendix 2. Composition of the Phillips Group[29] (per cent)

Phillips Petroleum Company Norway (USA)	36.96
Norske Fina A/S (Belgium)	30.00
Norsk Agip A/S (Italy)	13.04
	80.00
Norsk Hydro A/S (Norway)	6.700
Elf Norge A/S (France)	5.396
Aquitaine Norge A/S (France)	2.698
TOTAL Marine Norsk A/S (France)	4.047
Eurafrep Norge A/S (France)	0.456
Coparex Norge A/S (France)	0.399
Cofranord A/S (France)	0.304
	20.000

Appendix 3. STATOIL (per cent)

Frigg[30]		
Norsk Hydro A/S	32.870	(13.6)[a]
Elf Norge A/S (operator)	27.613	(38.4)
TOTAL Marine Norsk A/S	20.710	(28.8)
Aquitaine Norge A/S	13.807	(19.2)
STATOIL	5.000	
Heimdal[31]		
STATOIL	40.000	
Pan Ocean A/S	19.375	(36.905)[a]
K/S Femogtyvefire Norsk A/S	10.750	(20.476)
Bow Valley	8.000	(15.238)
Sunningdale Oils Norge A/S	3.875	(7.381)
Norsk Hydro A/S	6.228	(6.920)
Elf Norge A/S (operator)	5.2326	(5.814)
TOTAL Marine Norsk A/S	3.924	(4.360)
Aquitaine Norge A/S	2.6154	(2.906)
Statfjord[32]		
STATOIL	44.4423	
Mobil Development Norway A/S (operator)	13.3327	
Norske Conoco A/S	8.8885	
Esso Exploration Norway A/S	8,8885	
A/S Norske Shell	8,8885	
Saga Petroleum A/S	1,6665	
Amoco Norway A/S	0.9259	
Amerada Hess Norwegian Exploration A/S	0.9259	
Texas Eastern Norway A/S	0.9259	

[a]Figures in brackets denote shareholdings prior to STATOIL participation.

Appendix 4. Staff Growth and Education[33]

Growth	
1973	54 persons
1974	118 persons
1975	244 persons
1976	401 persons
1977	506 persons
1978	607 persons
1979	710 persons

Education (1976) (per cent)	
University educated engineers	31.2
Technical college engineers	7.3
University educated economists	12.2
Business college graduates	2.4
Geologists/geophysicists with university degrees	10.1
Lawyers	1.2
Secretaries	9.3
Others	26.3
	100.0

Appendix 5. The Location of Frigg, Heimdal, and Statfjord

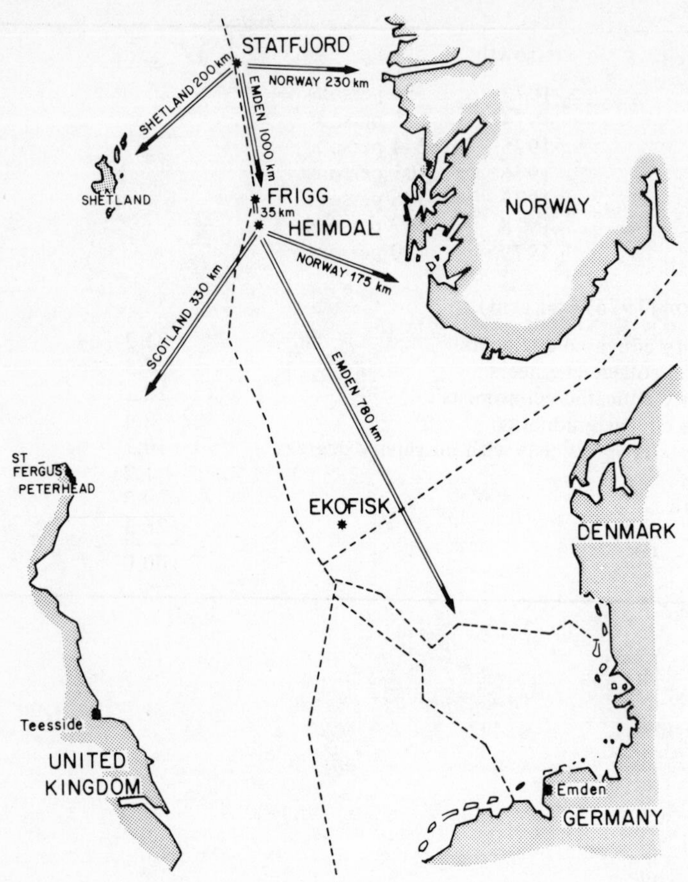

Appendix 6. Royalty and License Fees (per cent)

Royalty

1. For concessions 001–036 the following general rules have been applied:

Oil	10
Gas	12.5

As a consequence of later 'farm-in' agreements, some amendments have been made:

PL 013/014/015	Deminex/Pelican	10.5
PL 019/020	Conoco/BP/Pelican	12.5
PL 022	Murphy (half of concession oil)	5.0

Area Fees (Source: The Petroleum Directorate, *Annual Report*, 1975)

1. From April 9, 1975, the following amendments have been made for concessions 001–036:

Fees	Old (NKr per km^2)	New (NKr per km^2)
7 year	500,000	900,000
8 year	1,000,000	1,800,000
9 year	1,500,000	2,800,000
10 year	2,000,000	3,700,000
11 year	2,500,000	4,600,000
12 year	3,000,000	5,600,000
13 year	3,500,000	6,500,000
14 year	4,000,000	7,400,000
15 year	4,500,000	8,400,000
16 year onward	5,000,000	9,300,000

Reconnaissance Licenses

The fees for reconnaissance licenses have remained unchanged at NKr20,000 per annum to be paid in advance, for the duration of the license.

Appendix 7. Offshore Operations – Special Tax Provisions

By an act of June 13, 1975, new tax provisions were given with regard to exploration and exploitation of submarine natural deposits. The most important provisions are the following:

1. Norm Prices

The government may stipulate binding norm prices for petroleum products produced by offshore operations. The norm price will be applicable at the calculation of an ordinary tax, a special tax, and a royalty. The norm price shall be equivalent to the price at which petroleum could have been sold between independent parties in a free market. The appraisal shall take into account the realized and quoted prices for petroleum of the same or equivalent nature, making the necessary adjustments for differences in quality, transportation costs, delivery time, payment time, and other terms. The norm price will be stipulated by a special price committee appointed by the government.

2. Ordinary Taxes

With the following exceptions the ordinary tax provisions shall be used also by calculation of ordinary taxes for offshore activities:

(a) The gross income will be assessed according to the *norm price* (see above).
(b) The rate for ordinary depreciation shall be 16.66 per cent per year. No extraordinary depreciation.
(c) Deduction for operating losses may be claimed in the *fifteen* following years. However, the deduction in any year is limited to one-third of the loss in each of the preceding years.
(d) Losses and deficits from offshore operations are deductible, if derived from operations on the Norwegian Continental Shelf. Deduction of one-half of losses and deficits from *onshore* operations in Norway will be allowed.
(e) Income rates are the same as for other companies – at present 50.8 per cent.

3. Special Tax

This special tax will be applicable to companies which are engaged in production and pipeline transportation in offshore areas. The rate of the special tax will be stipulated each year by the Parliament. For 1975 the rate has been stipulated to 25 per cent. The special tax will be based on the income calculated for the purpose of ordinary income tax, but dividends paid will not be deductible, and losses or deficits from onshore activities will not be deductible. The special tax is payable on that part of taxable income which *exceeds* 10 per cent of the cost of those depreciable assets which have been acquired during the last 15 years. If 10 per cent of the cost of assets, as mentioned above, is higher than the net income calculated in accordance with the act, the excess amount may be charged as a deduction in later years for the purpose of assessing the special tax. The special tax is not deductible in calculating the ordinary tax.

4. Royalty

In addition to income taxes and net worth taxes a royalty is imposed on offshore operations based on gross value at the well-head. Originally, the royalty was 10 per cent. By a new Royal Decree of December 18, 1972, the royalty was set at from 8 to 16 per cent depending on the size of the production. For natural gas, the royalty was set at 12.5 per cent.

For an updated version see Thore Morch, 'Petroleum Taxation in Norway', *DNC Oil Now*, Number 4, pp. 6–29, Oslo. No date but estimated at 1978.

Appendix 8. Investments and Ownership

Investments[34] (millions of kroner)

Joint ventures	1976	1977	1978	1979	1980	Sub-total	Sum
Statfjord	984	1,693	3,476	3,250	3,056	12,459	
Frigg	181	89	52	45	35	402	
Heimdal	2	168	342	474	205	1,191	
Noretyl	240	14	–	20	–	274	
Norpolefin	193	195	93	42	–	523	
Rafinor	553	3	4	4	4	568	
Coast Center Base	5					5	
SUM	2,158	2,162	3,967	3,835	3,300		15,422
Corporation[a]							
Norpipe A/S	57	22	3			82	
Norpipe UK Ltd	13	4				17	
Norsk Olje A/S	92					92	
SUM	162	26	3			191	191
Directly administered							
Geological and geophysical explorations	41	39	47	43	35	205	
Base	7	6	1			14	
Research and development	3	4	4	3	3	17	
Administration building	27	48	51			126	
Other investments	19	11	37	6	5	78	
TOTAL	97	108	140	52	43		440

Ownership[35]

Company/license	Type of activity	STATOIL share (per cent)	Location
Statfjord	Oil/gas find	50	Blocks 33/9, 33/12
Heimdal	Gas find	40	Block 25/4
Frigg	Gas find	5	Block 25/1
STATOIL/Esso Group	Operator	50	Blocks 15/12, 15/11, 6/3

(Continued)

Appendix 8. (*Continued*)

I/S Noretyl	Petrochemicals	33	Bamble
I/S Norpolefin	Petrochemicals	33.33	Bamble
Coast Center Base Ltd. & Co	Supply base	50	Sotra
Norsk Olje A/S	Distribution comp.	15	Oslo
Rafinor A/S	Refinery	30	Mongstad
Norpipe A/S	Pipeline company	50	Stavanger
Norpipe Petroleum UK Ltd.	Landing terminal	50	Teesside

[a]The investments in the limited company are the payments of share capital (or the takeover of shares, in the case of Norsk Olje A/S).

Appendix 9. Income and Expenditure[36]

Balance sheet, 1973–75

Amounts (millions of kroner)	1975	1974	1973
Sales	382.3	98.2	–
Salaries and social insurance	20.1	8.0	2.1
Depreciation	1.1	0.3	0.2
Financial expenditures	7.9	7.0	7.6
Financial result	−62.2	−29.0	−13.7
Annual investments	956.8	65.3	238.3
Total assets	1491.6	502.9	359.6
Issued share capital December 31	755.0	305.0	155.0
Number of employees December 31	244	118	54

Expected operating results, 1976–80 (millions of kroner)

	1976	1977	1978	1979	1980
Crude oil and refined products	804	1,224	1,483	2,622	3,866
Gas	–	14	77	144	160
Petrochemical products	–	105	267	392	501
Dividends on Norpipe shares	–	17	49	60	60
Other income	24	12	10	10	11
	828	1,372	1,886	3,228	4,598
Operational expenses	823	1,363	1,779	2,680	3,088
Interest expenditures[a]	110	318	598	958	1,256
Depreciation	87	143	243	690	1,212
Net income before taxes	−192	−452	−734	−1,100	−958
Taxes	5	9	9	11	10
Net income after taxes	−197	−461	−713	−1,111	−963
Accumulated net income	−298	−759	−1,502	−2,613	−3,581

[a]For calculation purposes, interest is set at an average of 10% on all loans.

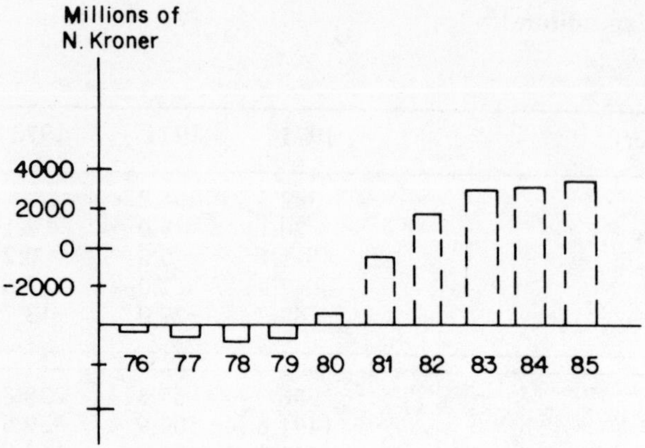

Contribution from operations from 1976 through 1985.

Appendix 10. STATOIL's Petroleum Directorate

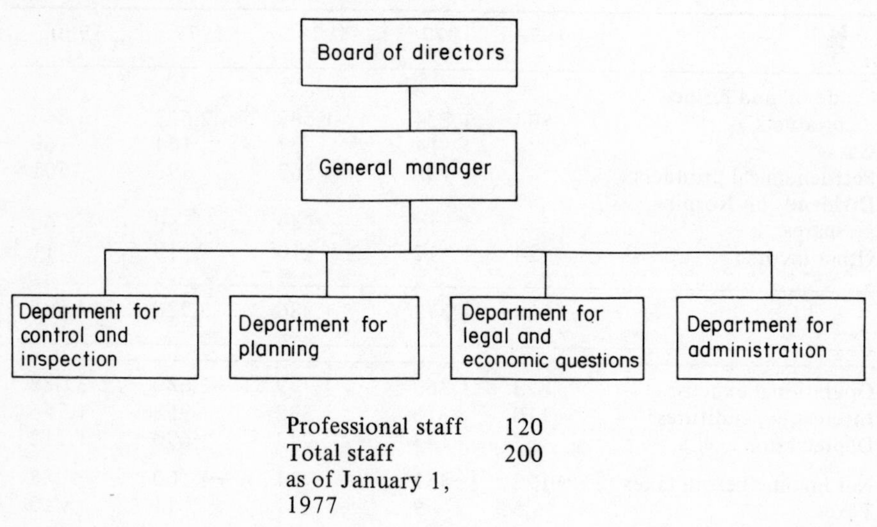

Professional staff	120
Total staff	200
as of January 1, 1977	

Appendix 11. Ministry of Industry – Department of Petroleum and Mining

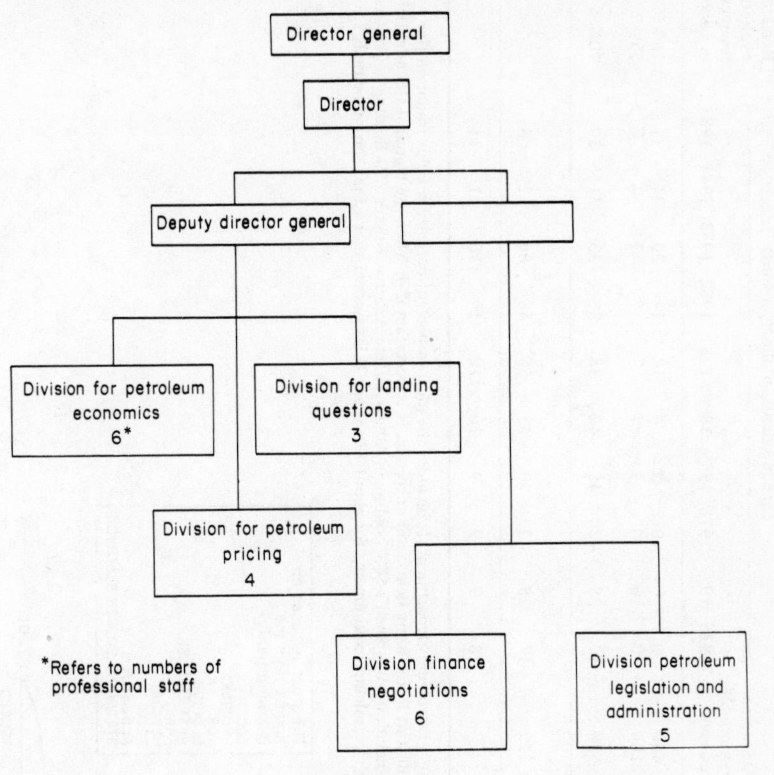

Appendix 12. STATOIL's Organization

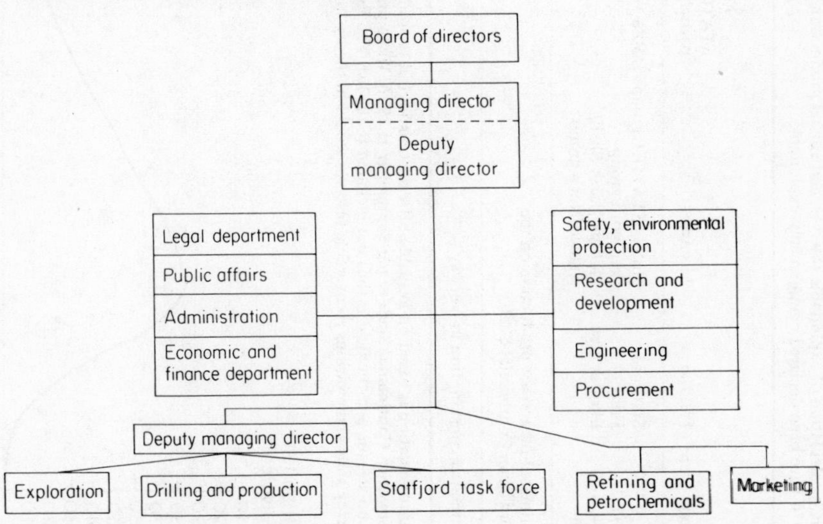

Appendix 13. STATOIL's Oil Reserves[37]

CnC's calculation of exploitable reserves, and annual production for the period 1976–85 for the oil- and gasfields on the Norwegian Continental Shelf, which as of September 1976 have been declared commercially exploitable.

Block	Field/area	Licensee	STATOIL's participation	Estimated exploitable reserves	Oil equivalents (millions of tons) 1976	1977	1978	1979	1980	1981	1982	1983	1984	1985	Remaining exploitable reserves in 1986
33/9–12	Statfjord	Mobil–STATOIL group	50%	440	0	0	0	2	6	10	14	20	24	31	76%
25/1	Frigg	Petronord group	5%	100	0	0	0	3	7	7	7	7	7	7	55%
2/4, 5–7 1/6	Ekofisk area	Phillips/Norsk Shell/ Amoco/Noco group	0%	570	17	28	42	45	42	38	35	33	31	30	40%
Estimated oil and gas production on the Norwegian Continental Shelf				1,110	17	28	42	50	55	55	56	60	62	68	
Estimated consumption in Norway					8	9	9	9	10	10	10	10	11	11	

This calculation is based on available official statistics but is produced by CnC's Petroleum Department. It includes the most important petroleum fields found on the Norwegian Continental Shelf as per September 1, 1976. When production will start is, for some fields, still uncertain. It is uncertain if or when the marginal Heimdal field will go on stream, and the field is thus not included in the list. The Ekofisk area includes Norsk Shell's 50% holding in Albuskjell and Amoco–Noco's 25% holding in Tor. As the survey shows, the various recoverable reserves would in theory meet Norway's petroleum requirements for about 130 years. It is assumed that 1,000 Nm^3 gas equals 1 ton oil.

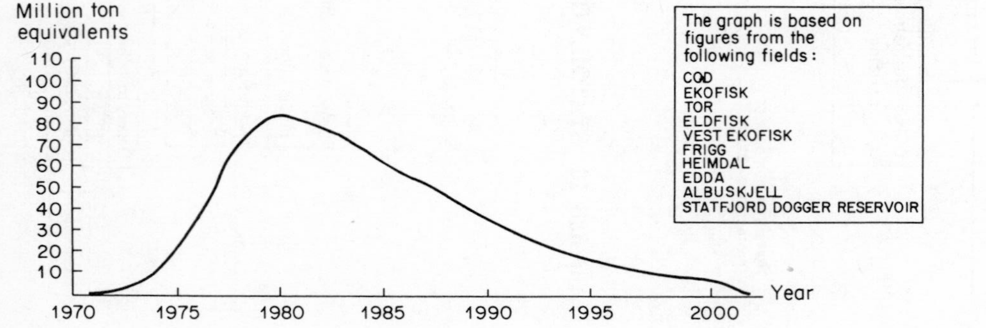

Appendix 14. Natural Gas Reserves

Norwegian Shelf, south of 62°N, 1976 (billions of cubic meters of gas)

	Billions of m³
Ekofisk area, associated gas	370
Frigg and Odin dry gas (Norwegian portion)	150
Statfjord, associated gas (Norwegian portion)	90
Sleipner, gas condensate	75
Heimdal, gas condensate	35
	720
Percentage of world reserves	1.1

Source: *Statoil*, January 1976.

References

1. *Operations on the Norwegian Continental Shelf*, Report 30 to the Norwegian Storting, 1973–74, Ministry of Industry, pp. 5–6.
2. Nils B. Gulnes, 'The Norwegian Petroleum Development', *Fifteenth Conference of the International Bar Association*, Vancouver, July–August 1974, p. 8.
3. Report 30 to the Norwegian Storting, 1973–74, Ministry of Industry, pp. 22–23.
4. *Landing of Petroleum from the Ekofisk Area*, Report 51 to the Norwegian Storting, 1972–73, Ministry of Industry, p. 6.
5. *Landing of Gas from the Frigg Area*, Report 77 to the Norwegian Storting, 1973–74, Ministry of Industry.
6. M. Kvinnsland, 'Aspects Economiques de la Politique Pétrolière Norvégienne', *Revue de l'Energie*, p. 130.
7. STATOIL, *Annual Report*, 1973.
8. 'Special: Offshore Gas', *Noroil*, April 1976.
9. STATOIL, *Annual Report*, 1973. For a downward revision, shown in the text, see *Oil Now*, No. 1, Den Norske Creditbank, Oslo, 1976, No. 1, p. 15.
10. Report 30 to the Norwegian Storting, 1973–74, Ministry of Industry, p. 44.
11. *Activities on the Shelf, of STATOIL and the Petroleum Directorate*, Fig. 2, Storting Proposition 81, 1974–75, p. 22.
12. Arve Johnsen, *STATOIL – Objectives and Perspectives*, Mimeographed, Stavanger, January 15, 1976.
13. *The Economist*, July 26, 1975.
14. Report 81 to the Norwegian Storting, 1973–74, Ministry of Industry.
15. Report 81 to the Norwegian Storting, 1973–74, Ministry of Industry.
16. Parliamentary Report No. 16.
17. Report 30 to the Norwegian Storting, 1973–74, Ministry of Industry.
18. Nils B. Gulnes, 'The Norwegian Petroleum Development', *Fifteenth Conference of the International Bar Association*, Vancouver, July–August 1974, pp. 3–7.
19. Private communication to the author, December 1976.
20. *The Economist*, July 26, 1975.
21. 'Disagreement on "Norm" Prices', *Petroleum Economist*, April 1976, p. 148. See also 'A Comparison of Tax Systems', *Petroleum Economist*, May 1976, pp. 170–172.
22. Report 21 to the Norwegian Storting, 1973–74, Tables 4 and 5.

23. Report 21 to the Norwegian Storting, 1973–74, paragraph 5.4.
24. *Petroleum Economist*, February 1976, July 1976, and August 1976.
25. 'Oil Brings Norway Riches and Problems', *New York Times*, April 20, 1978, pp. 49 and 61.
26. Private communication to the author, Spring 1978.
27. B. A. Gierde, 'The Fourth Round of Allocations in the Norwegian North Sea and Where Does Norway Go From Here?', *DNC Oil Now*, August 1979, pp. 1–10.
28. Norwegian Government, Oslo, No. 5-1979/80.
29. Report 51 to the Norwegian Storting, 1972–73, p. 54.
30. Report 77 to the Norwegian Storting, 1973–74, p. 6.
31. Report 77 to the Norwegian Storting, 1973–74, p. 7.
32. Storting Proposition No. 114, p. 3.
33. Private communication to the author, 1976.
34. Report 21 to the Norwegian Storting, 1973–74.
35. STATOIL, *Annual Report and Accounts*, 1975, p. 5.
36. Report 21 to the Norwegian Storting, 1973–74.
37. Den Norske Creditbank, *DNC Oil Now*, No. 1, p. 15.

CHAPTER 9

The Energy Policies of the European Community and the Role of the National Oil Companies Therein

The Treaty of Rome, establishing the predecessor institutions of the European Communities, went into effect in January 1957. At the time of this writing, it is more than five years since the 1973–74 quadrupling of energy prices created an 'energy crisis' for Europe and demonstrated the need for a common energy policy. But it is now clear that even by 1980 the member states of the Communities have not established a common energy policy. There are many reasons for this failure. This chapter will briefly review them, emphasizing the national oil companies' role in thwarting a common policy.

In 1957, responsibility for the EEC's energy policies was divided among three communities: coal was placed under the jurisdiction of the ECSC, nuclear power under Euratom, and oil and gas under the EEC. But treaties establishing these communities have not provided for the coordination of national energy policies. In this respect energy policy is unique; even such thorny problems as a common agricultural policy and a common transport policy were provided for in the EEC Treaty.

The omission is interesting. Depending on one's inclinations, one may credit negotiators for the EEC Treaty with foresight in avoiding an explosive problem or with irresponsibility in evading provisions for a coordinated energy policy. The problem of policy appeared anyway, and the fact that no arrangements had been made for its solution did not ease the situation.

The Interexecutive Working Party, comprising representatives of the EEC, the ECSC, and Euratom, was established in October 1959 to create a common energy policy. By the time that the first set of proposals was submitted in March 1960, the EEC energy economy had been transformed from one of post-war shortages to one of a serious surplus. The first proposal and the subsequent ones – all submitted throughout the sixties – had the joint objectives of assuring a desired level of EEC domestic energy production and, correspondingly, protecting the coal industry from the rapidly advancing petroleum imports.

During the sixties the EEC had only six members, and its debate about common policy may be simply rendered as a debate between the coal-haves and the coal-have-nots. Italy and the Netherlands generally opposed protection, because they had no significant coal industries. Belgium, Germany, and Luxemburg had significant coal industries and wished to protect them. France's coal industry was declining; however, France had two large national oil companies for which it had created an elaborate protective system. A common energy policy would threaten this systematic encouragement, and France proved more than reluctant to take any measure that would weaken its national oil companies. Italy, which also had a large national oil company, was equally reluctant to continue protecting coal at the expense of oil. In the coal-have countries, the coal industry represented significant vested economic, political, and social interests. In countries with national oil companies, these enterprises likewise represented significant vested economic and political interests. Both coal and oil were vital to national security. No EEC member state proved to be prepared to give up its sovereignty over such strategic commodities.

The purely economic dimension of the conflict between the 'haves' and the 'have-nots' can be expressed in terms of price. During the early sixties, Italy and the Netherlands – which were 'energy-poor' – had to import the major share of their fuels. They were, therefore, interested in any energy source as inexpensive as possible. Because Belgium, France, and Germany wished to protect their coal industries they favored higher prices for imported fuels. However, this description is oversimplified. Within each member state, there was a conflict between producing and consuming interests, the latter wanting low prices and the former not. Moreover, energy requirements of other industries caused further pricing conflicts among the member states. For instance, countries with rapidly growing basic metals and chemical industries favored low-cost energy, as fuel costs are significant in these industries. Countries with rapidly growing food and textile industries were less concerned about energy costs, as fuels represented only a fraction of total costs in these industries.

Besides the mainly economic reasons for the failure to agree on a common energy policy, there are reasons which more properly belong to the sphere of political economy. The member states, to a varying extent, were unwilling to give up sovereignty over their fuel policies and followed their separate national aspirations in this field. French energy policy may be summed up in a statement by the French Minister of Industry: 'We won't accept an open common market that leaves the security of Europe's energy supply in the hands of foreign companies.'[1] The 'foreign companies' indicated were American, British, or British/Dutch. (Britain was not then a member of the EEC.)

While French energy policy is the subject of Chapter 2, two special aspects need emphasis here. One is the Law of 1928, which established a state monopoly over petroleum. The other is the French preference for Franc Zone crude. Both the law and the policy succeeded in their intents to protect and support the two French national oil companies, CFP and the predecessors of SNEA.

Reviewing the early attempts to frame an EEC energy policy and to regulate the supply/demand situation during the sixties, one cannot help but share the query of the President of the ECSC: 'If the six countries could have foreseen that one day there would be not too little coal but too much, would they have drafted the [ECSC] Treaty as they did?'[2] The answer, of course, is that it is highly unlikely. Note that during the same period Europe was awash with inexpensive oil. The 'energy crisis' in the sixties consisted of a surplus of oil and coal.

The role of the NOCs in the European energy picture is a post-World War II phenomenon, for before 1945 the only NOC with a substantial crude oil base outside Europe (and in the Middle East) was CFP. The other NOCs were created after World War II, and even then the EEC was slow to realize that both NOCs and energy policy were important to the Community. Nor did the EEC countries take uniform steps to support NOCs and to establish national – as opposed to common – energy policies. France (see Chapter 2) regulated its market to encourage its NOCs and so did Italy, to a lesser extent (see Chapter 5). Both countries used their NOCs to encourage geographical diversification of oil supplies and direct cooperation with the OPEC countries. All other European countries relied on the oil MNCs to provide their oil supplies.

The EEC's failure to establish a common energy policy was aggravated by its failure to project the long-term economic and political implications of national energy policies. The myopia – or short-term expediency – may be summed up in an oft-repeated phrase of the fifties and sixties; the EEC wanted *inexpensive and secure* energy. Of course, the EEC should have realized that in the long run the policy tradeoff would become inexpensive *or* secure energy. Low-cost energy aided the rapid economic growth that politicians wished to promote. Only in 1973–74 did the EEC countries begin to realize the dual consequences of their dependence on OPEC oil: the disruption of supplies and the steep rise in prices.[3]

In 1967 the three communities were merged, and three new member states were admitted in 1972: the United Kingdom, Ireland, and Denmark. In 1968 the EEC still saw no conflict between inexpensiveness and security. And the EEC Commission could still assert 'the community's interests demand first and foremost security of supply at prices which are relatively stable and as low as possible. There is no foundation for the assertion that this requirement is a contradiction in terms.'[4] The Commission's first recognition of the problem came after the OPEC price rise began (in 1971), though before prices quadrupled in late 1973. In 1972 the Commission suggested that 'long-term security be given priority over temporary price advantages.'[5] The expansion of the EEC membership from six to nine did not facilitate agreement on energy. At the time of the EEC's expansion, British oil policy was heavily influenced by the oil MNCs, while French policy was influenced by its two NOCs. Thus, the British joined the Common Market and introduced a new element of conflict.[6]

The quadrupling of oil prices in 1973–74 should have finally brought about a common energy policy. The opposite turned out to be the case: energy now became a source of profound division within the EEC. Henri Simonet, EEC's Commissioner for Energy, acknowledged that the EEC exhibited an 'inability to face a major challenge in a way commensurate with its claim to be a major economic power evolving progressively into a political one.'[7]

The reasons for the post-1974 failure were both political and economic. Member states argued, for example, over EEC policy toward the United States and toward the Arab–Israeli conflict.[8] A common oil policy, it has been said, is '10 per cent oil and 90 per cent politics.'[9] As the pressure from OPEC increased, so did the 'politics' component of oil policy.

The chief economic factor behind the further politicization of energy issues after 1973–74 was the long-term transformation of the energy base of European economies from indigenous coal to imported oil. In the matter of imported oil, the role of the NOCs became crucially divisive. The French and the Italians supported their NOCs and the French treated their domestic oil markets in a *dirigiste* fashion, while Germany, the United Kingdom, and the Netherlands relied on oil MNCs and pursued largely *laissez faire* energy policies. Behind the *laissez faire* approach was the perception that political control over the globally oriented oil MNCs was obviously limited.

Coincident with the 1973–74 energy crisis was the general emergence of the British and Norwegian sectors of the North Sea as sources of oil for Europe. Alas, this favorable development also proved to be a divisive rather than a harmonizing influence. In early 1975, the EEC's President pointed to the dangers of 'the low road of intergovernmental oil cooperation [in energy policy] when we should be taking the high road of integration.'[10] This time it was, hardly surprisingly, the British who joined the French in their lack of enthusiasm for a common energy policy. Of the large industrial countries, the United Kingdom alone was potentially self-sufficient in energy. As a Parliamentary committee report proclaimed, 'of all the community members, we [the United Kingdom] alone, through the North Sea reserves, are in sight of self-sufficiency in oil, we have maintained a strong coal industry and we have natural gas. We have been pioneers in the development of nuclear power. No other member state can match these resources and experience.'[11] The United Kingdom by the mid-seventies was a large producer of oil as well as a consumer. Any attempt at establishing a common energy policy that would lessen the United Kingdom's control over the British sector of the North Sea – and to a lesser extent over the disposition of the oil produced in the British North Sea – was looked upon with disfavor by the UK government.[12] Nonetheless, in 1976 a British industrialist argued that although self-sufficiency during the eighties may make a purely national fuel policy attractive for Britain, it is in the long-term interest of the United Kingdom as well as the eight other member states to arrive at a coordinated EEC energy policy.[13] This contrary view did not sway British policy makers.

In addition to the United Kingdom, Germany and France have also altered their energy policies. Germany, in mid-1978, merged some of the oil and gas operations of Veba (the German NOC) and the German affiliate of BP (subject to approval of the cartel authorities).[14] (For details, see Chapter 6.) The merger had two consequences. First, it diluted the government's control over Veba, and, second, it made crude oil available to Veba, which has been critically short of oil. The merger came on the heels of some successes of Deminex, Veba's exploration and production company.[15] Deminex has a 41 per cent share in Thistle field in the British sector of the North Sea, a field which came on stream in mid-1978. In addition, Deminex announced a 'significant' discovery in the Gulf of Suez. The

German energy market, generally, remained 'free', and the German economy has exhibited slow growth rates since 1974. As a result, coal consumption continued to drop, notwithstanding the rather heavy subsidies this industry continued to receive. This decrease occurred in spite of the fact that the German government, while continuing its nuclear program, scaled down that program somewhat, expecting to use more indigenous coal in electricity generation.[16] Germany's heavy dependence on imported oil was partly mitigated by the dramatic rise of the Deutschmark in relation to the US dollar, which occurred between 1973 and 1979 when oil bills were being paid in the latter currency.

The seemingly major change has, however, occurred in France, where, following the electoral victory of President Giscard d'Estaing, some liberalization of the oil industry was announced. *Dirigism*, the name of the game since the late fifties, began to be phased out.[17] Import quotas and price controls will be lifted and marketing licenses will be granted more liberally. The latter move – given the excess refining capacity both in France and in Europe as a whole – will, undoubtedly, mean lower prices in France. However, the Law of 1928, which governed the French oil market, is not being scrapped. Companies will still be required to submit their import programs before obtaining their import licenses. The main purposes of de-control are to reduce prices to customers, thus helping to reduce the rate of inflation, and also (according to the government) to allow competition to reduce the oil account in the French balance of payments. (Oil in 1978 accounted for 20 per cent of all French imports.) It seems, though, that lower prices to consumers will increase consumption and, hence, oil imports. The government has not yet announced how it plans to cope with this conflict.

The impact of liberalization on the two French NOCs is difficult to predict. They prospered in the 'hothouse' environment of the Law of 1928. By 1978 they together held about half of the French market, and both were on their way to becoming multinational. In the government's view, liberalization implies that the two NOCs are capable of competing with the oil MNCs both at home and abroad. The chief executive officer of SNEA, Albin Chalandon, requested government assistance after liberalization was announced, but so far the government continues to feel that the two NOCs are strong enough to cope with the increased competition. If the government proves incorrect in its assessment, there is little doubt – given the history of fifty years of state intervention in the oil industry – that the French government will reverse itself and support its two NOCs.

It may seem paradoxical in the late seventies to assert that the EEC's continued failure to arrive at a common energy policy is caused not by too little energy, but by too much – i.e. by excess refining capacities throughout Europe, which put a considerable strain on the 'European' companies. In September 1976, the five 'European' companies – CFP, SNEA, ENI, Veba, and Petrofina – requested the EEC to forbid all oil companies to grant discounts (see Chapter 2). The EEC has not acted on this request. The problem of excess capacity did not improve, prompting the president of CFP to note the merciless price wars raging on the Continent while 'Europe of the Common Market has nothing to say for itself.'[18] In view of the French attitude toward a common energy policy, a strange comment indeed!

That oil policy continued to be 10 per cent oil and 90 per cent politics was confirmed by EEC inaction on energy matters between 1974 and 1978, the year of the latest set of meaningless proclamations. In late 1974 and early 1975 the EEC Council 'affirmed its political will to draw up and implement a Community energy policy.' The resolution went on to state that such policy would enable the EEC 'to express a common viewpoint on energy problems *vis-à-vis* the outside world.'[19] No actions followed these affirmations. The EEC Commission also adopted a set of 'aims for 1985',[20] urging rational use of energy, greater use of internal and secure resources, greater use of nuclear energy, and increased research and development and alternative (nonconventional) energy sources. Again, the EEC Commission did not indicate how the Community could get from the state of affairs of the mid-seventies to the 'aims of 1985'. In the sphere of international action, the EEC deferred to the newly established International Energy Agency (IEA). However, France refused to join the IEA. The EEC has cast itself as an observer, but the IEA did little, if any, international policy coordination.

Undaunted, the EEC Council of Ministers, in December 1975, urged the member states toward 'solidarity in the case of tight oil supply, to encourage energy saving and to favor the development of EEC energy resources.'[21] What has been said of the earlier attempts also applies to the guidelines of December 1975.

Yet another effort of the EEC to arrive at a common energy policy was foiled in mid-1978[22] when the member states rejected proposals dealing with excess refining capacity and subsidies for coal. The United Kingdom supported the coal plan and rejected the proposals dealing with refineries. Italy would not accept the coal proposal unless the refinery issue was also solved. Besides the United Kingdom, Germany also supported the coal subsidy program, while France, a large importer of coal, joined Italy in its primary interest in the excess refining problem.

It must be apparent from the review of the last twenty years that it is highly unlikely that a common energy policy in the EEC will emerge in the foreseeable future, if ever. European governments' support for their national oil companies has proven to be yet another obstacle to a common energy policy. Yet some kind of energy policy must be pursued. After all, the lack of a policy is itself a policy – the policy of doing nothing.

References

1. Maurice Bokanowski quoted in *Petroleum Intelligence Weekly*, New York, January 18, 1963.
2. Piero Malvestiti, *Sources of Energy and Industrial Revolutions*, The European Communities, Luxemburg, 1961, p. 25.
3. Guy de Carmoy, *Energy for Europe*, American Enterprise Institute, Washington, 1977, pp. 37–40.
4. 'First Guidelines for a Community Energy Policy', Supplement to *Bulletin of the European Communities*, No. 12, 1968, p. 7.
5. 'Necessary Progress in Community Energy Policy', Supplement to *Bulletin of the European Communities*, No. 11, 1972, p. 15.

6. 'How Nine Energetic Men Took 19 Hours to Disagree', *The Economist*, May 26, 1973.
7. Henri Simonet, 'Energy and the Future of Europe', *Foreign Affairs*, April 1975, p. 452.
8. Robert J. Lieber, *Oil and the Middle East War: Europe in the Energy Crisis*, Center for International Affairs, Harvard University, Cambridge, 1976, pp. 44–46.
9. Henri Simonet quoted by Martin Mauthner, 'The Politics of Energy', *European Community*, March 1974, p. 13.
10. *Eighth General Report (1974) of the EEC*, Brussels, February 1975.
11. *Membership of the European Community – Report on Renegotiation*, Cmnd. 6003, p. 33.
12. *Twenty-Second Report on the European Communities*, Session 1974–75, R/3333/74/75, HMSO, London. For an update, see Department of Energy (Energy Paper Number 22), *Energy Policy Review*, published by HM Stationery Office, London, 1977.
13. Richard Bailey, 'Headings for EEC Common Energy Policy', *Energy Policy*, December 1976.
14. *Petroleum Economist*, July 1978.
15. *Petroleum Economist*, June 1978.
16. *Petroleum Economist*, December 1977.
17. Doris Leblond, 'Changing the Rules for Oil', *Petroleum Economist*, October 1978.
18. CFP, *Annual Report*, for 1977, Paris.
19. *Official Journal of the European Communities*, July 9, 1975.
20. *The European Community and the Energy Problem*, European Documentation, Brussels, 1975/2.
21. European Community, *Background Information*, Washington, January 29, 1976.
22. 'No Agreement on Energy Policy', *Petroleum Economist*, July 1978.

CHAPTER 10

Policy Instruments to Promote National Oil Companies

Chapter 1 asserted that national oil companies cannot be launched, nor survive, nor prosper without government assistance. The NOCs' very *raison d'être* demands such assistance. For although the NOCs' objectives are partly financial, the companies have also been set up to perform a set of sociopolitical and economic services that, in the view of their owner governments, the oil MNCs cannot adequately provide. A second reason that NOCs need government support is that the international, vertically integrated oil industry posed certain barriers to entry. Barriers included the MNCs' sheer size, their access to diversified and relatively inexpensive crude supplies, and their worldwide marketing networks. In addition to these advantages, the oil MNCs had much 'old' investment (i.e. pre-high inflation), while the NOCs had to start or expand heavily in the inflationary economies of the seventies. In these circumstances, it is rational for governments to subsidize investments that otherwise would not be so supported.

To promote their national oil companies, the five countries under review here have set up quotas, granted exploration licenses, allocated rights to service stations, fixed prices, used NOCs in government to government dealings, and backed NOCs in the international money markets.

We might expect that government assistance should be easy to identify; after all, one should be able to recognize a subsidy when one encounters it. Alas, the realities are much more complicated. It is not altogether clear whether some of the measures are deliberate government assistance, or whether some of their results are actually only economic rent. I hope to analyze the policy measures and their results in such a way as to allow the reader to make up his or her mind whether a particular measure represents government support or the sound business practice of a prudent owner.

Ownership and Capital

Of the six NOCs under review, three are fully state owned: ENI, STATOIL, and BNOC. The French government owns 35 per cent of CFP (and holds 40 per cent of CFP's voting rights) and owns 70 per cent of SNEA. The German government owns about 44 per cent of Veba. NOCs that manage to expand and invest without any further government support after the initial subscription to capital are comparable to private enterprises whose owners provide funds for investment and working capital either through equity or debt. Alternately, if either return on investment (ROI) or cash flow is insufficient and the government does inject further capital to support the NOC, this support may or may not be comparable to that extended by a conglomerate to one of its inadequately performing companies, or by a multidivisional firm to one of its enterprises.

Several aspects of Veba's financing illustrate the difficulty of making such comparisons. The German government gave Veba DM 140 million, the sum exceeding the market value of Gelsenberg shares, to entice owners of the latter company to exchange their shares for those of Veba. This transaction is quite similar to a tender offer that one company extends to another – regardless of ownership – if it wants to acquire the company for which the tender offer is made. On the other hand, a clear example of governmental subsidy is offered by Deminex, Veba's exploration and production company. In this instance, the German government subsidized Deminex to the tune of DM1,375 million between 1969 and 1978 for exploration outside Germany. A further DM600 million has been made available to Deminex for the years 1979–81. In addition to these direct subsidies, preferential loan and repayment schedules were also arranged between the government and Deminex. Assistance particularly difficult to classify is the subsidy that Veba received for its coal production through Veba's 27.2 per cent ownership of the consolidated Ruhrkohle AG. In 1975, for instance, Ruhrkohle received an average subsidy of DM23 per ton on coal produced. However, the majority private owners of Ruhrkohle received the same subsidy from the government. Hence, this measure cannot be said to favor specifically the German NOC, Veba.

ENI's capital structure also illustrates the ambiguity of the phrase 'governmental assistance'. ENI's initial capitalization consisted of state holdings (comparable to owner's equity) and debentures (owner's debt), plus an assessment of income from its petroleum exploration in Italy on behalf of the state. However, much of the exploration had been conducted by ENI's predecessors. In addition to the initial capitalization, ENI received huge amounts of reinvested state profits between 1954 and the present. In some sense these state reinvestments are comparable to the retained earnings of any corporation, private or public. But the analogy should not be carried too far. ENI's owner, the state, has immediately reinvested all dividends, a phenomenon unlikely to occur in any private corporation over a span of twenty-five years.

Direct financial subsidization of a NOC can be best illustrated by French support for SNEA. The French government has used four different tax systems that benefited SNEA directly or indirectly. In 1950 the government introduced a petroleum products sales tax to subsidize national oil production. Virtually all the proceeds went to ERAP, the predecessor production company of SNEA. The tax was phased out in 1973. Proceeds of an earlier sales tax, later merged into the 1950 tax, financed the Institut Française de Pétrole (IFP), which was set up in 1943 for hydrocarbon research. Beginning in the late fifties, IFP became increasingly involved in exploration, production, and refining. Theoretically, IFP's results were available to CFP and to those oil MNCs that operated in France, as well as to SNEA. However, the former were going concerns that relied on their own research facilities. Thus, SNEA benefited disproportionately from IFP's work. Two temporary taxes were enacted in the wake of the 1973–74 'oil crisis'. A 'tax parafiscale', enacted in March 1974, penalized companies with relatively large market shares in petrol (the oil MNCs) and favored those with disproportionately large shares in fuel oil (primarily SNEA). Finally, an inventory reserve tax had a one-time detrimental effect on the oil MNCs, which in 1974 recorded large inventory profits. Because the charter of SNEA (and CFP) did not allow them to register inventory profits, these companies escaped the tax. The proceeds of this tax were also, in a large measure, recycled to SNEA.

Quotas

In addition to granting capital and revenue to their national oil companies, countries set up import barriers to protect their NOCs. The most widely applied nontariff barrier is the quota, which puts a quantitative limit on the imports allowed into a country each year. One of the main reasons for using quotas rather than tariffs is that quotas give the government authority and flexibility to achieve certain goals. Laws that grant the government power to issue quotas also give officials discretionary authority to issue import licenses and to determine their duration.

France used the quota system extensively over at least half a century to establish and to assist its two NOCs, CFP and SNEA. The Law of 1928 specifically set out plans to develop a French oil industry. The quotas then issued continue, in modified forms, to the present. Crude import quotas (A20) were awarded for twenty years, and finished product quotas for petrol and lubricants (A3) were valid for three years. The crude import quotas were designed to protect refiners, the product quotas to protect marketing companies. In the original A20 quotas, French groups received almost 55 per cent of the licenses. Over the next fifty years, French groups received around half of the total quotas. To give the government greater flexibility, the time span of the crude import quotas was shortened – in 1951, to thirteen years; in 1965, to ten years. At present, in the 1976–85 period, French companies have 55.5 per cent of the quotas.

The principle of A3 quotas of 1928 was to promote the building of refineries in France by prohibiting companies from being net importers of finished products. A 1952 ordinance obligated the refiner–marketers to refine 90 per cent of their sales in France. In 1960, one of SNEA's predecessor companies directly entered the refining–marketing phase of the business and obtained quotas for the first time. As a result of the A3s, the oil MNCs' market share dropped from 80.5 per cent in 1945 to 60.4 per cent in 1955 and to 47.3 per cent in 1975. In the 1928–31 period, the government also gave CFR, CFP's refining company, the right to refine 25 per cent of French consumption. This privilege, coupled with the provision that 90 per cent of all product sales had to be refined in France, gave CFR a tremendous boost in its refinery program.

Under the Law of 1928, the government had the additional power to obligate refiners to buy crude from specific countries. In the early and mid-fifties sizeable quantities of crude were discovered in the Franc Zone; in allocating quotas, the government gave first preference to this crude. Because the Franc Zone crude was more expensive than that of the traditional suppliers (primarily the Persian Gulf countries), and because the oil MNCs had relatively little Franc Zone crude, this arrangement strengthened the NOCs, particularly SNEA. Most of the Franc Zone crude owned by SNEA came from Algeria. The government priced the crude at such high levels that the obligation to buy Algerian crude became an indirect subsidy to SNEA.

While the oil MNCs lost market share to the NOCs in France, their absolute sales increased because the size of the market grew dramatically. Also, because the quota system prevented new companies from entering the market, the oil MNCs' operations in refining and marketing over much of the last fifty years have been more profitable in France than in those European countries where a free market existed. Thus, with a certain degree of governmental approval, the major companies completed 'industry agreements' on three of the major products. Such agreement was based essentially on the A3 quotas. The agreements broke up in 1964 because of the government-supported entry of an SNEA company into marketing.

Provisions of the French Law of 1928 ran counter to the Common Market Treaty – or so its Commission maintained. Throughout the sixties and most of the seventies only marginal concessions were made to the Common Market partners, and the quota system remained in effect without any substantial change. The French quota system was finally 'liberalized' in late 1978 – not because of Common Market claims, but because the world oil situation had radically altered, moving from persistent surpluses to sizeable future shortages. Whether the liberalization was real or illusory remains an open question. In the wake of the Iranian oil shutdown in the spring of 1979, the French Minister of Energy, André Giraud, justified France's quota system thus: 'Free market forces cannot be applied in a tight market.'

The United Kingdom has also set up a quota to try to restructure its refining industry. In establishing BNOC, the government insisted that 66 per cent of British North Sea production be refined in the United Kingdom. The oil MNCs were equally insistent that they refine only 35 per cent in the United Kingdom. Thus, while the French government-owned refining industry has indeed been built up by consistent application of the quota system, the British quota system is still under debate.

Exploration and Production Licenses

Access to inexpensive crude is the *sine qua non* of the oil industry. Because barriers to entering the exploration and production phases are high, all five governments under review here assist their NOCs' backward integration with licensing agreements favoring the NOCs. Governments also grant licenses to oil MNCs or to other foreign private companies on condition that they in turn grant equity participation to the licensing country's NOC. Thus, preferential licensing agreements allow NOCs low-cost entries into exploration and production. Conversely, equity participation through licenses granted to MNCs admits the national oil companies to essentially no-risk entries. The issuing governments find that such licenses are less expensive and less politically risky than full direct ownership of exploration firms.

Two countries whose NOCs were short on crude, France and Italy, have used licensing arrangements to try to balance crude supplies with national needs. France's CFP acquired its crude oil in the Persian Gulf by extension of force of arms. Prior to World War I, the Deutsche Bank held 25 per cent of the Turkish Petroleum Company, predecessor of the present Iraq Petroleum Company (IPC). After the war, through the San Remo Treaty of 1921, the victorious French state obtained this quarter share in IPC from the Germans. Oil was found on the concession in 1927. Ever since, Iraq has been first the exclusive and subsequently the major crude source for CFP.

In 1924, the French government set up an organization which gave public companies and private companies incorporated in France exclusive exploration rights, both in France and in overseas French territories. As a result of this encouragement, gas was found in Southwest France before World War II. Oil was found after the war in various overseas French territories, notably in Algeria, Gabon, and the Congo (for details, see Chapter 2).

In 1926, the Italian government gave exploration rights in Italy and abroad to AGIP, ENI's predecessor company. After World War II, the government granted ENI onshore and offshore exploration monopolies for both oil and gas. Large quantities of natural gas have since been discovered in the Po Valley but little oil has been found.

For the two countries long on crude, Britain and Norway, issuing exploration and production licenses became the primary means of launching national oil companies. It is unlikely that either BNOC or STATOIL would have been founded had oil and gas not been discovered in the North Sea. The first licenses were awarded in the United Kingdom in 1964–65. Government policy favored British-based companies and companies that had refining and marketing operations in Britain. All licenses were awarded to private companies; of these, 33 per cent went to British companies and 45 per cent to American companies – most with substantial UK operations. In 1976, BNOC was founded. In the licensing round of 1977, the condition of obtaining an exploration license was granting 51 per cent participation to BNOC. After much protest from the private companies, the licensing round was successful. Some sixty-five companies – including all the oil MNCs – applied for and were given licenses, and all licensees granted BNOC 51 per cent participation. Moreover, the government gave preference to those companies that made BNOC participation retroactive to licenses granted between 1964 and 1977. All private companies – including the oil MNCs – granted BNOC retroactive participation.

Norway issued its first licenses in 1965, and state participation became the condition for the second round of licenses, awarded in 1969. The first significant find was made in 1969. When STATOIL was established in 1972, the company's participation varied from a 5 per cent interest in an early field to a 50 per cent share in the huge Statfjord field. The licensing conditions were amended in 1975 to give STATOIL a minimum interest of 50 per cent in any new concession – an interest that could rise to 75 per cent if the find was substantial.

Providing Access to Preferential Marketing Outlets

Providing access to crude oil has been a major form of governmental assistance. Promoting the NOCs' entry into markets hitherto controlled primarily by the oil MNCs has been another. Both the French and the Italian governments gave preferential treatment to their NOCs in setting up service stations. The French system was rather subtle. Under the guise of physical planning, the government gave some preference to Elf, SNEA's marketing company, by awarding service station licenses along the *autoroute* system. In Italy, beginning in the early seventies, ENI was awarded 50 per cent of the new service station licenses along the *autostradas*. These high volume outlets in both France and Italy had above-average profitability.

Price Fixing of Oil and Gas Production

Economics is primarily concerned with the *efficiency* with which society's resources – or those of a firm or household – are used. Some issues, however, deal with *equity* – i.e. with the distribution of income. A particularly vexing equity problem is caused by fixed quantities of resources – in this case oil and gas. At any given price, the supply of oil and gas is inelastic. Those individuals, firms, and countries that have found oil or gas on their lands stand to collect what economists call 'economic rent'. Economic rent *per se* is not governmental assistance. It becomes such assistance if it is allocated in a discriminatory manner. The firm owning the oil and/or gas may collect economic rent through its market power, or it may have the rent assigned to it, as it were, by the government. One method of governmental alloction of economic rent is price-fixing. Price fixing, coupled with the monopoly rights certain governments gave their NOCs began methods of assistance. Nowhere, however, are the lines clearly cut. In the case of the MNCs, is it an 'economic rent' which is being collected or do the benefits come from their market powers and/or early entries into the market?

ENI illustrates the ambiguity of this point. ENI obtained exclusive exploration, production, and transportation rights for oil and gas for the entire Po Valley. Moreover, ENI has consistently priced the gas – which was produced in large quantities – in relation to fuel oil rather than production costs. ENI's gas profits were crucial to its survival, because they underwrote ENI's marketing expenses. With the quadrupling of oil prices in 1973–74 and the subsequent increases since that period, ENI raised its gas prices to keep them in line with the equivalent fuel prices refined from OPEC crude. Similarly, SNEA (as well as its predecessor company, SNPA) has priced its gas found in Southwest France at the world market fuel oil prices and not at production costs – again showing fairly large production profits. Of course, all companies, private or public, have correlated gas prices to fuel oil prices – regardless of production costs. For instance, the oil MNCs had obtained considerable profits for decades on their oil production in the producing countries. These production profits cross-subsidized their downstream activities just as ENI's gas profits financed its other activities.

The case of Veba is a counter-illustration. Germany has some domestic oil and sizeable gas production. It so happens that the great bulk of that production is in the hands of four oil MNCs – Esso, Shell, Mobil, and Texaco. Veba, the German NOC, has no domestic production of either oil or gas. The persistent increase in crude oil prices made large gas production profits for the four oil MNCs. These profits cross-subsidized downstream activities as did the gas profits made by ENI and SNEA. Veba's lack of domestic oil and gas and the corresponding lack of production profits made any subsidization of its marketing operations impossible. Thus, the government's policy of not taxing domestic 'windfall' profits put Veba at a competitive disadvantage *vis-à-vis* the four oil MNCs.

The cases of BNOC and STATOIL are considerably simpler. Every time OPEC raises crude oil prices, the comparable prices demanded by BNOC and STATOIL increase correspondingly.

Price-Fixing of Product Prices

Price-fixing at the product phase, which was done in both France and Italy, proved to be of uncertain assistance to their NOCs. Whether price controls helped or hindered the NOCs depended on their market shares. As long as their market shares were low – and the corresponding shares of the oil MNCs were large – the burden fell primarily on the latter companies. However, as the NOCs' market shares increased – and in France the NOCs' combined share now exceed 50 per cent – the burden imposed by price controls increased correspondingly. Similarly, when ENI acquired Shell's assets in Italy in 1974 and increased its market share from 25 per cent to 40 per cent, the marketing losses that ENI began to realize became a concern to ENI management just as they have been to the oil MNCs and CFP and SNEA. Theoretically, an NOC was in a better position than the oil MNCs to ride out the losses, although the extent to which the NOCs were able to finance these losses depended on their crude and gas profits and on their market shares. As the latter increased, executives of CFP, SNEA, and ENI joined their oil MNC counterparts in both France and Italy pleading for relaxation of controls on product prices.

NOCs as Instruments in Government to Government Dealings

When in 1973–74 not only were oil prices raised but also oil supplies from certain OPEC countries curtailed, some governments got into the act of scrambling for supplies. In 1974 the German government obtained rights to twelve million tons of

Saudi crude over the three years of 1975–77, designating Veba as the recipient of that crude. While the price was apparently the going market price at the time, availability in itself was a means of assisting the crude-short Veba.

ENI made particularly good use of government to government dealings, starting as early as 1957 in Iran and Egypt. In the sixties ENI concluded numerous deals with governments of certain less developed countries where, in exchange for crude or for building refineries, the Italian government and ENI became involved in arms sales, technology transfers, barter arrangements, and a host of joint investments. In 1972, following the IPC troubles with the Iraq government, ENI negotiated a supply arrangement with the Iraq government for 20 million tons of crude over a ten-year period. In these deals, the Italian government and ENI were indistinguishable: the government seemed to be an instrument of ENI's foreign oil policy.

The French government has used both CFP and SNEA to negotiate government to government deals. Iraq granted CFP a ten-year supply contract; CFP is also a shareholder in IPC. However, most of France's oil negotiations were conducted through SNEA. In the late sixties Iraq and SNEA concluded a service agreement whereby SNEA received a supply contract in exchange for sharing its exploration and production expertise. In July 1979, the French government announced that Iraq would increase oil sales to SNEA from 175 million barrels in 1979 to 210 million in 1980. In return, the French government has pledged to deliver the Osirak nuclear reactor to Iraq in 1982. As a result of this agreement, Iraq is expected to supply one-third of France's oil imports in 1980, thereby becoming France's largest supplier after Saudi Arabia. SNEA has also made supply arrangements with Iran and Mexico. In the mid-seventies, Iran agreed to supply oil in exchange for SNEA's production technology. During President Giscard's 1979 visit to Mexico, that country agreed to provide SNEA with 100,000 barrels per day of crude in exchange for French goods and services. Also in 1979, SNEA gained a foothold in the People's Republic of China, although no details of that arrangement have been released.

Government Backing of NOC Loans

The six NOCs under review in this book are sufficiently established to be able to borrow both in their domestic and in the international capital markets. Nonetheless, government backing affords them easier access to these markets and probably better borrowing terms as well.

Preferentially treated NOCs were ENI and Veba's exploration and production company, Deminex. In Italy, ENI had favored access to the domestic banking system. The nature of this access is not altogether clear, except that ENI has been able to borrow at times and under circumstances when private companies would not have been able to. Also, throughout the seventies ENI successfully raised funds in the Eurocurrency markets. Beginning in 1974, ENI's annual income statements and balance sheets became progressively less promising, and at the same time the Italian government was also in a sorry financial state. That ENI was continuing to raise funds in the Eurocurrency markets was due mainly to the liquidity of that market rather than to ENI's financial viability or that of its guarantor, the Italian government.

Germany's Deminex estimated that its development costs would reach DM7 billion by 1982. Most of this outlay has been and will continue to be financed through loans – with the help of government guarantees. Deminex's financing is somewhat similar to that of BNOC and STATOIL in that the loans are primarily to support exploration and production. However, the government's role as guarantor is different in all three cases. In June 1977, BNOC raised $825 million in the Eurocurrency markets without explicit government backing, but obviously the lenders were aware of BNOC's 100 per cent public ownership. Prior to 1975, STATOIL had been able to borrow directly, a right that was then revoked. Since that time, the government has borrowed on STATOIL's behalf and, after scrutinizing STATOIL's capital requirements, has turned the appropriate funds over to STATOIL. Nonetheless, both BNOC and STATOIL own such volumes of oil and gas that there is little doubt that either of them could raise domestic and foreign exchange funds in the capital markets without government guarantees.

Governments have assisted their NOCs in a variety of ways and over an extended period of time, although it is difficult to classify such assistance in an industry characterized by market imperfections. Chapters 3 through 8 describe in more detail the extent of the support and its duration. As of 1980, BNOC and STATOIL are awash with cash, and CFP and SNEA are financially independent of the French government. Except for efforts to subsidize Deminex, the German government has never really supported Veba, and because of Veba's partial merger with BP (Germany) the company's financial future is unclear. ENI was financially self-sufficient during the first two decades of existence. Beginning in the mid-seventies, however, as ENI's domestic gas supplies shrank, so did the company's ability to generate cash. It is likely, therefore, that in the future ENI will become more dependent on government assistance than it has been in the past.

CHAPTER 11

The Relationship Between the National Oil Companies and the Oil Multinationals

The exploration and production phases of the oil industry have long been characterized by joint ventures. Collaborating companies can share the cost of capital, share technology, and share political risk. It is therefore not surprising that the 'Big Four' oil producing countries in the Persian Gulf – Saudi Arabia, Iran, Iraq, and Kuwait – all had concessions operated by joint ventures.

In European downstream operations, the primary focus of this chapter, the oil MNCs preferred to retain their control by retaining 100 per cent ownership. US anti-trust laws also hindered joint ventures among American-based companies. The French, German, and Italian NOCs, on the other hand, freely entered into joint ventures. There are several reasons for their involvement. First, the NOCs are smaller than the oil MNCs and (with the exception of CFP) are latecomers to the oil industry. Thus they have less available capital and management expertise. Opportunities to share capital and technology are therefore attractive. Second, companies long on crude can share supplies with firms short on crude, an arrangement that saves both parties from incremental investments in refining and marketing, and also protects downstream operators from additional competition. Far from being hampered by anti-trust laws, joint ventures are encouraged in European countries. In fact, they are all pervasive among French and Italian domestic and international operations.

Downstream operations in the less developed countries have also taken the form of joint ventures. In the sixties, several less developed countries insisted that refineries be built within their borders. But markets in most of these countries were too small to support even one refinery. Thus, most of these refineries are joint ventures, the participating companies' shareholdings usually in proportion to their domestic market shares.

This chapter will discuss joint ventures between the national oil companies and the multinationals. I will use the term *joint venture* to refer to an enterprise in which at least two of the partners are oil companies, including long-term supply contracts as well as ventures in which equity and management are joint. Long-term supply contracts have always been an important feature of the oil industry, since some of the largest NOCs were short on crude while most of the MNCs were long on it. As this chapter will demonstrate, such supply arrangements have been critical for some of the NOCs, particularly ENI, SNEA, and Veba.

This chapter concentrates on joint ventures in refining and marketing in Europe. Were I to deal with exploration and production ventures in any detail, the discussion would become unnecessarily cumbersome. For example, through licensing agreements, BNOC and STATOIL are engaged in joint production ventures with every oil MNC, every NOC, and virtually every independent. Including refineries outside Europe would similarly cloud the central relationships among the oil companies, since the jointly owned refineries in less developed countries bring CFP, SNEA, and ENI into joint ventures with virtually every oil MNC, with one another, and with several independent companies.

CFP and the BP Connection

CFP did not need to arrange joint ventures or long-term contracts to obtain a supply of crude. The San Remo Treaty of 1921 provided this supply, by granting CFP at 23.75 per cent interest in the Iraq Petroleum Company (IPC) as reparation from the Germans after World War I. CFP's partners were British Petroleum (BP – the original holder of the concession), Shell, and the jointly owned affiliate of Exxon and Mobil. Each of the major partners held a 23.75 per cent interest; the remaining 5 per cent was held by Gulbenkian interests. CFP's presence in the Middle East eventually opened additional doors. In the early fifties, when the Iranian Oil Participation was renegotiated, CFP received a 6 per cent interest. The two largest shareholders were BP (40 per cent) and Shell (14 per cent). The rest of the Iranian Participation was held by the five American oil MNCs, each with 7 per cent, and by a group of American independents holding a total of 5 per cent.

CFP's interest in IPC began its association with a group of oil MNCs – and particularly with BP – which had been adept at finding low-cost oil. In the early fifties, CFP, mostly through the MNCs that operated IPC, discovered and began to produce crude in Qatar, Abu Dhabi, Oman, and Dubai. CFP's association with BP extended to the French market when, in 1955, the two companies took joint ownerships in one of the largest independent French marketers. This venture helped to establish CFP as the major marketing company in France.

CFP's other principal partner was, understandably, SNEA. The two companies became partners in an exploration and production company in Algeria. The two companies continued their joint exploration efforts in the British and particularly in the Norwegian sectors of the North Sea, with considerable success.

These ventures eventually led the two companies to some joint enterprises in petrochemicals and marketing. In 1969, prompted by the government, CFP and SNEA affiliates entered a 50–50 enterprise in petrochemicals, ATO Chemie and Compagnie de Petrochemie. The companies' joint marketing also involved two American oil MNCs – Standard Oil of California and Texaco: the largest remaining French independent marketer, ANTAR, was taken over in 1970 by SNEA (41 per cent) and CFP (24 per cent); Caltex, jointly owned by Standard of California and Texaco, took 20 per cent, the state owned 10 per cent, and the original owner kept 5 per cent.

SNEA and the Caltex Connection

SNEA's predecessor companies in exploration and production were frequently involved in joint ventures. As indicated above, CFP later shared some Algerian ventures with SNEA. Significantly, however, so did the second largest oil MNC, Shell, as well as a host of American independents. Shell's participation in Algeria was logical; it held a sizeable market share in France and it was one of the oil MNCs short of crude.

Once oil was found in both North and West Africa, SNEA needed downstream operations to find captive outlets for its crude. Caltex, whose shareholding companies were long on crude and who had a small and not very profitable position in France, recognized the potential business synergy. In the early sixties, SNEA took a 60 per cent interest in the Caltex refinery at Bordeaux (the original owner retained 40 per cent) and took full ownership of Caltex's French distribution outlets, thus buying an immediate 4 per cent market position. In 1974 the SNEA–Caltex arrangement was renegotiated. Caltex's ownership was transferred to a 24 per cent interest in Elf-France (the French refining and marketing subsidiary) and renegotiated once again, after the merger establishing SNEA in 1976 gave Caltex a 19 per cent interest in SNEA. Also in 1976, Caltex undertook a long-term supply arrangement with SNEA – the latter still short of crude – to supply it annually with 4 million tons of Persian Gulf crude at preferential prices on a long-term basis. Caltex representatives were and are members of the SNEA Board of Directors.

One of the SNEA marketing companies is the independently managed ANTAR, acquired in 1970. As mentioned above, ANTAR's principal owners are SNEA, CFP, and Caltex, the latter holding 20 per cent of equity and supplying the same proportion of crude. ANTAR holds about 9 per cent of the French market.

ENI and the Exxon (Esso) Connection

Mattei, ENI's founder and first chairman, scornfully called the oil MNCs the 'Seven Sisters'. He became even more incensed at the oil MNCs when ENI found itself excluded from the newly restructured Iranian Oil Participation in 1953–54. Exxon, as the largest oil MNC, ought to have been a special target for Mattei's ire. Yet, paradoxically, ENI and Exxon enjoyed a long and mostly amicable relationship.

One of ENI's original refining companies, STANIC, was equally owned by ENI and Esso. Esso had a long-term supply arrangement for that refinery, providing crude oil not only for its own but also for ENI's account. (Another early ENI refinery was owned 51 per cent by ENI and 49 per cent by BP, with a similar long-term supply arrangement for crude.) As the fifties drew to a close and ENI was still short of crude, Mattei approached Esso with the idea of a larger long-term supply contract. At that time there was a glut of oil, and Esso proved to be an eager partner. In 1963, shortly after Mattei's death, the two companies signed a five-year, 10 million ton agreement. The arrangement was unusual, inasmuch as Esso agreed to take half of its payments in machinery and services from assorted ENI enterprises. Esso was not only long on oil, it was long on gas as well. In the mid-sixties, Esso and ENI signed a twenty-year natural gas supply contract of 3 billion cubic meters per year for Libyan gas – to be transported in refrigerated tankers. In the early seventies, ENI signed a long-term supply agreement for Dutch gas, the ownership of which was held by the Dutch government, Esso, and Shell. It is estimated that Esso is involved in nearly 40 per cent of ENI's gas imports.

ENI had to turn to other oil MNCs for crude as it continued to be chronically short of it. In 1964, ENI took over Gulf's small oilfields in Sicily and also obtained a supply arrangement of 5 million tons over a six-year period at substantially below posted prices. To facilitate the arrangement, the Mellon family, who were Gulf's substantial shareholders and who had large banking interests, made a preferential loan arrangement. Another oil MNC from which ENI obtained crude was Shell. As part of the deal to sell its downstream operations, Shell undertook to supply ENI with 21 million tons of crude for five years, beginning in 1974. ENI was a 10 per cent partner with a host of oil MNCs in the Trans-Alpine pipeline which carried crude from Trieste to Southern Germany.

Phillips, an American independent oil company, was a recurring and successful partner in ENI's explorations outside Italy and in its petrochemical plants. ENI entered into exploration agreements with Phillips in the North Sea and jointly they made the first discovery in the Norwegian sector. (Other partners in the consortium were CFP and SNEA.) Phillips and ENI also had a joint exploration venture in Nigeria. In Italy, in the mid-fifties, ENI built a large petrochemical complex with Phillips' technical assistance.

Veba and the Mobil and BP Connections

Veba's involvement with Mobil is of long standing; its more substantial involvement with BP is more recent. Prior to World War II, Mobil bought a 28 per cent interest in ARAL, Veba's marketing outlet for petrol, diesel, and lubricants. Veba now holds a 56 per cent interest in ARAL; the rest is held by a German independent, Wintershall. Mobil never acquired its own retail outlets in Germany. In a long-term contract beginning after World War II and running through 1980, Mobil has supplied Veba with 3.5 million tons per year of crude at preferential prices. Mobil has been one of the largest German domestic producers of crude. However, overall German production has been relatively small and, therefore, most of Mobil's oil to ARAL has come from the Persian Gulf. Veba and Mobil were also joint owners of a refinery at Neustadt.

Veba's BP connection is more recent and was made final only in 1979. Since 1975 both Veba and BP have registered losses in Germany. Veba's main weakness was its lack of crude oil, and BP's weakness in Germany was its inadequate representation in refining and retail outlets. With worldwide operations, however, BP was profitable overall, while Veba's oil operations continued to be unprofitable. In the 1979 deal, BP and Veba signed a twenty-year contract (1980 through 2000) for BP to supply Veba with 3 million tons per year of crude at prices comparable to those by which BP (London) supplied crude to BP (Germany). In exchange, BP acquired an option to buy 1,000 petrol stations from Veba, stations that Veba had purchased from Gulf and that were not integrated into ARAL. In addition, BP acquired 50 per cent of Veba's refinery at Ingolstadt, Veba's 25 per cent share in the Speyer refinery, Veba's fuel oil and coal distribution company, Stinnes, and a 31 per cent interest in Veba's liquid natural gas import terminal at Wilhelmshaven. BP (Germany) concluded an agreement with Algeria to import liquid natural gas at 4.5 billion cubic meters per year over a twenty-year period. Under a complicated arrangement, after objections from the Cartel Office were overruled by the Economics Minister, BP also acquired Veba's 25 per cent share in Ruhrgas, but agreed that its interest would have nonvoting status. Veba obtained DM800 million in cash. Hence, Veba and BP have joined in a long-term oil-supply arrangement, in the importation of liquid natural gas, in refining, and in marketing fuel oil and coal.

STATOIL and the BP Connection

STATOIL, with its huge crude supply, began to search for direct outlets in Norway. (Almost 90 per cent of STATOIL's crude is destined for export.) Such an opportunity arose in 1976 when BP offered its Norwegian refining and marketing operations for sale to STATOIL's owner, the state. The purchase was made, and a new company, NOROL, was formed under private and state ownership. NOROL and associated companies own the second largest of Norway's three refineries and have acquired BP's 25 per cent market share in the domestic market. The sale may have been facilitated by two factors: BP (Norway) already had 50 per cent private Norwegian ownership, and was reported to be less profitable than its two private competitors in Norway, Esso and Shell.

BNOC and What Is The BP, Esso, and Shell Connection?

After BNOC was formed in 1976, the British government pressured all holders of concessions in the British sector of the North Sea to grant BNOC a 51 per cent participation. The pressure was probably most strongly felt by BP, Esso, and Shell, which held among themselves 55–70 per cent of the domestic market share and 45–60 per cent of the North Sea discovered reserves. In mid-1976 BP not only agreed to BNOC's participation but also undertook to train BNOC staff in its domestic refining and marketing operations. Similar agreements were concluded with Esso and Shell in early 1977. In these later agreements the two oil MNCs retained title to their North Sea oils in exchange for the same training provisions that BP had granted and they agreed to provide training in UK planning operations as well. By early 1977, all companies operating in the British North Sea signed similar agreements with BNOC. These contracts were made under Britain's Labour government; the Conservative Party is less committed to BNOC.

It seems that some BNOC participation in domestic refining and marketing is inevitable, but the Conservative victory in May 1979 makes the timing and the extent of this participation uncertain. If the Conservative Party decides to expand BNOC's participation, the framework for such expansion has been laid by the BP, Esso, and Shell agreements.

British Petroleum: A Special Case

All oil MNCs participated with the NOCs in joint ventures and long-term supply arrangements, but BP is more prominently involved than the other MNCs. The casual observer might suspect that the British government is responsible for the degree of BP's participation. The British government owns 48 per cent of BP and has two representatives on the board. This ownership share is larger than that of the French government in CFP and the German government in Veba. However, I do not believe that government ownership accounts for BP's pattern of joint ventures and agreements with NOCs. In fact, BP has acted as an oil MNC since its foundation in 1908, and the company has been so regarded by its part-owner government, by other oil companies, and by the general public. BP is one of the Seven Sisters. (That is why I have excluded BP from consideration as a national oil company in this book.) Several other factors account for BP's involvement in joint ventures. First, BP is one of the oldest MNCs and was the first to discover oil in the Persian Gulf. Thus, BP has had a history of dealing with governments. I believe that BP has been able to transfer its experience in dealings with governments of producing countries to its negotiations with governments of consuming countries. A more important reason for BP's joint ventures is purely economic. BP's traditional business strengths have been exploration and production – a dual expertise still evident in BP's current production successes in Alaska and the North Sea. BP is long on crude. Conversely, BP has traditionally been weak in refining and marketing. It made, therefore, perfect economic sense for the company either to provide crude on a long-term basis or to acquire large downstream operations, as the company has through its deals with Veba.

Rhetoric and Reality

While de Gaulle and Mattei heaped scorn on the multinationals, the French and Italian NOCs did business with Standard of California, Texaco, and Esso. Esso and Texaco, in turn, condemned governmental interference in the oil industry – interference which did not prevent these firms from doing business with European national oil companies.

These intracompany relations were only natural. Many of the MNCs entered into joint ventures with NOCs for the same reason that BP did – their need for distribution outlets. Although in 1980 it is difficult to grasp that the oil industry was glutted with crude up till the early seventies, the early entrants – the oil MNCs – had crude oil in excess. The latecomers – the NOCs – muscled into some of the markets hitherto held by the MNCs. Joint ventures among NOCs and

MNCs were also natural in the less developed countries, where companies shared the political and financial risks of exploration. And it was only business sense for companies to share capital, management, and technology in exploring the North Sea and other new oilfields. Finally, by engaging in joint ventures, the European national oil companies were moving toward the goal of becoming multinational themselves, for the same reasons that the Seven Sisters had become multinational – to spread the risks and to escape any one government's control over their operations.

CHAPTER 12

Management Policies and Techniques of National Oil Companies

In Chapter 1, I outlined the national oil companies' purposes, both financial and nonfinancial. Many nonfinancial objectives focused on domestic development. Thus, NOCs were charged with securing preferential access to vital raw materials, oil and gas; alleviating inflation; maintaining full employment; and aiding underdeveloped regions through local investments and employment. Extrinsic nonfinancial goals included improving the balance of payments; freeing the home country from foreign political and economic domination, in part by achieving technological superiority and economies of scale; and, in general, enhancing the country's international prestige. Most of these goals require substantial capital and human resources, and yield low returns on investment (ROI). Such objectives conflict with the NOCs' key financial directive: to maximize profits (or to earn 'adequate' ROI).

Conflicts among mutually exclusive objectives can never be resolved satisfactorily. In the absence of clear priorities, the manager's task is to balance competing claims. Management techniques in national oil companies offer consummate studies in the art of decision making in organizations with multiple aims.

Profits, Growth, or Influence?

Conventional wisdom has it that the aim of private business is to maximize profits. Serious doubts about this accepted wisdom were expressed as early as the thirties by Berle and Means, and with renewed vigor in the sixties and seventies by Baumol, Marris, and Galbraith. According to the Berle and Means–Galbraith theory, the goal of big business is twofold: growth and control. These economists suggest not that profits are ignored, but that if growth and profits conflict, the former prevails. The impetus for growth is very strong in the oil industry – the world's 'Biggest Business' – regardless of ownership. Growth is required not only to realize economies of scale, but to exercise control over the social, political, and economic environment in which the firm operates. The assumption here – a plausible one – is that the larger the firm, the better its ability to influence its external environment. There are also internal pressures to grow. The larger the firm the greater the power, prestige, salary, promotion opportunities, and job security available to the firm's executives and managers. There are, of course, constraints on the extent to which big business can grow and expand its influence, at the expense of profits. In the private sector the constraints are revolts of stockholders and creditors. With no clear-cut measure of 'maximum profits', management's goal is to generate enough profits to keep stockholders and creditors reasonably happy.

However, several factors prevent such constraints from hampering growth of publicly held companies. Obviously, if oil MNCs were to expand too rapidly, thereby earning inadequate returns on investment, they would have to either retrench or face dire financial consequences. National oil companies, on the other hand, could secure additional debt or equity from their owner governments, on terms not available to private companies. (Of course, this kind of financing also has its limits.) While all big corporations try to expand, NOCs can make growth itself their top priority. In this context, the rapid expansion of NOCs that we have observed in Chapters 3 through 8 can be appreciated as rational business behavior.

Furthermore, owner governments often prize the economic and political power that NOCs can create for their owners, above the profits that the NOCs generate. Hence, governments tend to react favorably to requests for capital from the NOCs, again reinforcing the NOCs' natural tendency to grow. Conversely, the more powerful the government bureaucracy, the greater a manager's commitment to sales growth that will enhance his firm's leverage with that bureaucracy.

Finally, one may safely assume that the primary motive of governments is to stay in power, using all public enterprises, including NOCs, to ensure re-election. Thus, NOCs are used to alleviate social problems, via regional investments and increased employment, for example. These functions in themselves promote the NOCs' further growth. In the scramble for public funds to support growth, NOCs have done well on the whole, primarily because they have satisfied such social objectives better than most other public enterprises, and because the NOCs – certainly by the standards of public enterprise – have been profitable, as Chapters 3 through 8 have shown.

Decision Making in Big Business – Risk Preferences

Whereas oil MNCs are overseen by their managers and owners, NOCs must satisfy ministries, parliaments, and the general public as well. And whereas both MNCs and NOCs are subject to constant surveillance by the news media, the owners of NOCs are more likely to be influenced by public opinion. Indeed, parliaments and media usually emphasize failures and ignore successes – particularly where public enterprise is concerned. Owners and managers anxious to avoid adverse publicity become cautious. Thus, while big business tends to be risk averse, national oil companies tend to be even more so.

Furthermore, a multiplicity of objectives often paralyzes innovation. Since governments rarely set clear priorities among these competing aims, NOC managers must evaluate the tradeoffs. Such evaluation is itself risky, for if a manager upsets a vested interest, he may simultaneously thwart his own career. This danger enhances the standard managerial urge to cover one's tracks.

Finally, the NOCs' nonfinancial goals engage these companies in high-risk, low-ROI activities that oil MNCs would not touch. Since ROI is not the sole criterion by which managers are judged and since governmental fiscal controls are influenced by political and social considerations, NOCs have, on occasion, undertaken some risky ventures. But NOCs have proven reluctant to undertake these projects and have done so only after explicit governmental assurances that failure would not be blamed on management. Notwithstanding these assurances, NOC managers frequently received the blame anyway – which made them even more risk averse.

Nonetheless, some NOCs have occasionally proved more enterprising than the oil MNCs. Also, the entrepreneurial 'first generation' chief executive officers were often more courageous than their successors. Among the companies under review here, five NOCs were put on the map by a new, dynamic breed of entrepreneurs: de Metz of GFP, Guillaumat of SNEA, Mattei and Cefis of ENI, Johnsen of STATOIL, and Lords Kearton and Balogh of BNOC. These were executives who combined business acumen with considerable political connections. They could afford to take risks, and did so. STATOIL and BNOC are so new that they are still operating with their first-generation chief executives. The other NOCs are now under managements considerably more cautious than their founder managements.

In the long run, managers' risk preferences are influenced by the way in which their performances are evaluated by ministers and parliaments. Because these measures are not exclusively financial, managers are less likely to be swayed in their decision making by purely financial considerations than by their reading of the critical balance among the governments' many financial and nonfinancial expectations.

The Means of Governmental Control of the NOCs

The most obvious means of exercising control is the power of the purse, and the ministries of finance have traditionally been the governments' fiscal watchdogs. Financial controls assumed varied forms. Governments can limit the availability of funds for either capital projects or working capital or both; they can make funds available in the form of debt rather than equity; and they can insist that NOCs pay adequate dividends to the government owners. All of these measures hamper managerial flexibility.

Formal controls are often less successful. In France control resides in the Ministry of Finance. It is my judgment that control has been inadequate. Fiscal controllers simply check whether NOC funds have been properly accounted for, not whether resources have been effectively used. No market comparisons are made regarding budgets, prices, costs, or margins. Control in Italy is even less effectively performed by the Court of Accounts, which is responsible to the Parliament. The controllers are lawyers who have little background or interest in economics and management, let alone entrepreneurship. Moreover, most controllers are from Southern Italy, while ENI managers are economists and engineers from the North. In both France and Italy the problem is compounded by the fact that NOCs are under the jurisdiction of more than one Ministry. (This is true for all NOCs.) Invariably the Minister of Finance is involved; the 'other' Ministries are Energy in France, State Holdings in Italy, Economics in Germany, Energy in the United Kingdom, and Petroleum in Norway.

Yet another means of governmental control is the political appointment of chief executives. The first and present chief executive of STATOIL is a political appointee, but one appointment is no basis for predicting a future pattern. In France, Germany, and the United Kingdom, chief executive appointments went to people well connected to the 'establishment' but without any specific party affiliations. A recent departure from this practice in France may be the appointment of Albin Chalandon as chief executive of SNEA in 1977. Chalandon, a former minister, has close ties to President Giscard d'Estaing. Nonetheless, the practice of political appointments has been consistently applied only in Italy. In the last twenty-five years ENI has had six chief executives, and all were affiliated with the Christian Democratic Party. This pattern ended in 1979 when the Christian Democrats, forced to form a coalition government, divided the three top public corporation positions among the Christian Democrats, the Socialists, and the Social Democrats. Under this scheme the ENI top position went to a Socialist. But it seems that even under this system managerial capability has been a prerequisite for the appointment. Party affiliation in ENI has been likened to marriage with the boss's daughter or membership in the 'old boy' network in private enterprise – it merits special consideration, but in the final analysis appointments will be based on managerial performance and promise.

Decision Making in Big Business: Desire for Autonomy

As in all big business, the primary goal of NOC managements is to reduce the owners' influence and to increase one's own managerial independence. NOC managers particularly want the authority to make the necessary tradeoffs among the competing goals charged to the NOC. One means of obtaining such power is to monopolize a technology by hiring all of the country's experts. Thus, CFP, SNEA, ENI, and BNOC have appropriated the expertise needed for exploration and production in the North Sea. ENI and SNEA have cornered the expertise needed for drilling and transporting natural gas; the Italian and French governments lack these capabilities. Only the Norwegian government has acquired a significant body of technological knowledge.

In addition to securing technological independence, NOCs seek financial independence from their owner governments. Earning a 'reasonable' return on investment fosters independence in several ways. First, retained earnings provide internal funds for investment. Second, a profitable NOC can obtain loans in the capital markets. Finally, ministries, parliaments, and the news media are less apt to scrutinize profitable NOCs than unprofitable ones.

A 'reasonable' ROI is tricky to achieve, however. If returns are too high, owner governments assume (1) that the NOC charges outrageous prices, (2) that it pays inadequate wages, (3) that it ignores some or most of its nonfinancial objectives, or (4) that it commits all three sins simultaneously. In fact, a national oil company's public position resembles that of the multinationals. The moment that ROI seems more than 'reasonable', governments and the news media begin to talk about 'excess' or 'windfall' profits. A very profitable NOC would lose some of its managerial independence – the government might ask it to undertake a new nonfinancial project, for instance. On the other hand, running the NOC at a loss would also erode managerial authority. In some cases, a breakeven income statement or a positive cash flow is sufficient to keep governmental influence at a minimum.*

A third means of keeping owner governments at arm's length is diversification. All four of the established NOCs are vertically integrated, STATOIL is partially integrated, and BNOC is becoming vertically integrated. But horizontal diversification is more likely to increase managers' independence, for the more diverse the businesses, the greater is the discretionary authority that falls to NOC managers. Diversification also makes financial comparisons with other oil companies less meaningful, thereby loosening governmental control and exaggerating managerial *anomie* among NOCs. Hence SNEA set out deliberately to diversify and did so into high-ROI ventures. Veba started as a conglomerate and still is one. ENI diversified partly by choice and partly under duress – the Italian government saddled it with insolvent enterprises. ENI did not want many of these, as they made 'reasonable' ROI more elusive. Nonetheless, these enterprises have increased the number of ENI's nonfinancial projects and thus the company's discretionary authority. The classic diversification case is SNEA. This company transformed itself from a single-product company to a truly diversified one, at the same time increasing both profitability and leverage with its government.

Going multinational is the final technique of gaining independence from the government. Foreign crude sources and markets cannot be controlled by home governments and can be influenced with great difficulty. That is why BNOC and STATOIL, whose crude sources are within national boundaries, are subject to stricter governmental control than the other four NOCs, most of whose crude sources are outside their home countries. Equally difficult to control are joint ventures and long-term supply arrangements when the 'other' party to the agreement is foreign. There are also some technical problems of control: for instance, governmental auditors have difficulty checking the books of the NOCs' foreign subsidiaries. Owner governments may prefer not to tamper with foreign subsidiaries, which can earn foreign exchange, an important consideration if the foreign exchange is scarce, as was occasionally the case in post-war Italy. More will be said on this subject in the final chapter.

In all big businesses, managerial discretion is considerable, as owners of large

*Relative de-emphasis of ROI is not restricted to the public sector. Harold Geneen made ITT's primary corporate goal an annual increase in earnings per share rather than ROI.

firms are widely dispersed. As long as most of the owners are content, managers of private big business have considerable authority. NOCs, however, have one owner, the government, and usually the government is an active owner. One would therefore expect that governments have extensive control over their NOCs and that private owners have only marginal control, if any, over the oil MNCs. This expectation is generally correct, but the extent of governmental control depends in part on the personality of the NOC's chief executive officer. As indicated earlier, the newcomers, STATOIL and BNOC, are relatively tightly controlled by their governments. At the other end of the spectrum are SNEA and ENI. Guillaumat of SNEA managed to marshall the government to its support. Mattei and Cefis of ENI controlled the ruling Christian Democratic Party, rather than its controlling ENI.

Resolution of Conflicts Between Governments and NOCs

NOCs are both public enterprises and oil companies; it is the mix of the two identities that creates conflicts between NOCs and their governments. The mix is dynamic; most NOCs started out more like public enterprises and eventually became more like oil companies. Initially, governments have a clear set of objectives for their NOCs, and the latter are dependent on governmental assistance. As NOCs become operational, jurisdictions become unclear, and conflicting claims are resolved through economic and political bargaining. Finally, when NOCs have become going enterprises and governmental assistance is no longer crucial, NOC managers try to shake off governmental control, deemphasizing their NOCs' identities as public enterprises and joining the ranks of multinational oil companies. Of course, this pattern of development does not fit all NOCs under all circumstances. ENI until 1976 told the Italian government what to do, for all practical purposes; SNEA and the French government worked hand-in-glove until 1977. But by now ENI and SNEA have joined STATOIL and BNOC in the general opposition between national oil companies and owner governments. Although the pattern varies in each case, all national oil companies illustrate the impetus in big business to secure managerial independence from the owners.

References

When planning this book I looked forward with excitement to this chapter. However, as the time came to commit pencil to paper I found that most of my thoughts were anecdotal. For instance, I remember asking an executive of the French affiliate of an oil MNC the difference between the managements of CFP and SNEA, and he replied that if the French government were to ask SNEA to build a service station on the second level of the Eiffel Tower it would do so, but CFP would refuse a similar request. Subsequent interviews, and my own experiences, convinced me that the situation is much, much more complex. This chapter is based more on my own interpretation of these interviews rather than on a set of facts that can be documented. Nonetheless, there is some literature available on the subject of management of public sector corporations, some more useful than others when applied to NOCs. I list the ones that helped my thinking.

1. *A Study of U.K. Nationalized Industries*, London, 1976, and *Nationalized Industries*, London, 1978.
2. Yair Aharoni, Zvi Maimon, and Eli Segev, 'Performance and Autonomy in Organizations: Determining Dominant Environmental Components', *Management Science*, May 1978.
3. Guy Arnold, *Britain's Oil*, Hamish Hamilton, London, 1978.
4. William J. Baumol, *Business Behavior, Value, and Growth*, Macmillan, New York, 1959.
5. Adolf A. Berle and Gardiner C. Means, *The Modern Corporation and Private Property*, Macmillan, New York, 1934.
6. John Blair, *The Control of Oil*, Pantheon Books, New York, 1976.
7. Kalman J. Cohen and Richard M. Cyert, *Theory of the Firm: Resource Allocation in a Market Economy*, 2nd ed., Prentice-Hall, Englewood Cliffs, N.J., 1975.
8. John Kenneth Galbraith, *The New Industrial State*, 2nd ed. revised, Houghton Mifflin Company, Boston, 1971.
9. John Kenneth Galbraith, *Economics and The Public Purpose*, Houghton Mifflin Company, Boston, 1973.
10. Harvey Leibenstein, 'The Missing Link – Micro–Micro Theory?', *Journal of Economic Literature*, June 1979.
11. Robin Marris, *The Economic Theory of Managerial Capitalism*, Basic Books, New York, 1968.
12. Herbert A. Simon, 'Rationality as Process and a Product of Thought', *American Economic Review*, May 1978.
13. Christopher Tugendhat and Adrian Hamilton, *Oil – The Biggest Business*, Methuen, London, 1975.
14. Kenneth D. Walters and R. Joseph Monsen, 'State-Owned Business Abroad: New Competitive Threat', *Harvard Business Review*, March–April 1979.
15. Kenneth D. Walters and R. Joseph Monsen, 'The Nationalized Firm: The Politicians' Free Lunch?', *Columbia Journal of World Business*, Spring 1977.
16. Oliver E. Williamson, *Markets and Hierarchies: Analysis and Antitrust Implications: A Study in the Economics of Internal Organization*, Free Press, New York, 1975.

CHAPTER 13

Will the Seven Sisters Become Thirteen Sisters?

This chapter contends that, if the NOCs are financially successful, after a period of time they will no longer meet the original objectives of the governments that established them. Governmental objectives will continue to be met only while the NOCs earn low returns. The NOCs' primary means of evading nonfinancial objectives is to join the ranks of the oil MNCs. The process is already well under way in four of the NOCs under review here, and it is likely to begin in the fifth. Only Veba as yet shows no signs of multinationalization. As this chapter will demonstrate, the process of national oil companies' becoming state-owned multinationals is determined by the nature of the oil industry and by the nature, certainly in Europe, of big business itself.

Multinationalization of state-owned companies is not exactly new – not even in such nationalistic countries as France. Aerospatiale, the French aircraft manufacturer, joined West German, Dutch, and Spanish aircraft manufacturers to produce the Airbus, in competition with Boeing, Lockheed, and McDonnell Douglas. Snecma, the French aircraft engine manufacturer, also joined with foreign manufacturers and signed a cooperating production agreement with General Electric. Renault, the second largest company in France (after CFP), forged a production and marketing link with American Motors and has European subsidiaries outside France and in Latin America. In the computer industry, Machines Bull first teamed up with General Electric, and the combined company merged with other firms in the French computer industry to form Compagnie Internationale pour l'Informatique (CII). This combined enterprise, however, did not prove commercially viable, and in 1975 CII merged with Honeywell to salvage its 'national' computer program. All of these French enterprises are owned by the state.

But several aspects of the oil industry make it a 'natural' for multinationalization. One is the geographic quirk that oil is essentially found where it is not consumed (the OPEC countries) and consumed where it is not produced – in Europe, Japan, and, in the last decade, the United States. There have been some exceptions, such as the United States until the late sixties and, as of 1980, the United Kingdom.

Second, as Chapters 11 and 12 have described, NOC managements try to spread the risks of doing business and to increase their own control over operations, at the expense of that of their owner governments. Vertical integration and diversification can accomplish both aims. Multinationalization extends the NOC's economic power internationally, just as vertical integration extends power over all phases of the oil business, and diversification extends economic power over activities not necessarily related to the oil industry but usually within national boundaries.

This pattern of expanding economic power was first followed by the Seven Sisters themselves. The original MNCs were Esso, Shell, and, somewhat later, BP. The other four oil MNCs were relative newcomers – Mobil, Texaco, Standard of California, and Gulf. Mobil was the only one of these who had 'ties' in IPC with the original oil MNCs. The other three latecomers followed the route of joint ventures and long-term supply arrangements, both with the established multinationals and with each other. Standard of California discovered in Saudi Arabia large crude reserves for which it had no overseas markets. Texaco, with a strong marketing position in the eastern hemisphere but with limited overseas crude supplies, joined Socal in 1936, thereby forming Aramco (for exploration and production) and Caltex (for refining and marketing). In the late forties, Aramco production exceeded Caltex requirements, so Aramco was expanded to include a 30 per cent share for Exxon and 10 per cent for Mobil, Socal and Texaco retaining 60 per cent. The pattern of long-term contracts was adopted in Kuwait as well, where, again in the late forties, crude production exceeded the requirements of the 50 per cent owner, Gulf (BP owned the other half). Thus, in 1950 Gulf signed a long-term supply agreement with Shell, which was short of crude, for specified amounts until 2026.

The so-called independent companies, Amoco, Arco, Conoco, and others, likewise took the path of multinationalization. For instance, Conoco established marketing outlets in the United Kingdom, Germany, Belgium, and Austria in the fifties. To provide crude for these outlets Conoco joined Marathon and Amarada in a company called Oasis, which by the mid-sixties exported 50 per cent of Libya's production.

I make this historical review to suggest that what makes business sense for the oil MNCs and the independents makes sense for the NOCs as well. As a matter of fact, the history of most oligopolies shows that worldwide interests become more important as the oligopoly matures. The oil industry is simply a prime example of this trend. Hence, to survive or thrive without continued governmental assistance, all NOCs will become multinational, thereby spreading risks, supplies, markets, capital, technology, and management. This trend necessarily has eroded the control that 'home' governments are able to exercise over the their NOCs, as the latter have to operate independently of their governments whenever national and multinational interests clashed. The timing of multinationalization has not, of course, been uniform. Older NOCs, such as CFP, are more likely to have acted as members of the oligopoly for some time; the newer ones, say BNOC, still act as agents of the state, but signs of multinationalization are already evident.

NOCs' activities outside their home countries are detailed in Chapters 3 through 8; by summarizing these activities here, I wish to highlight my point. In 1978, CFP sold 57 per cent of its total refined products outside France, both in and outside Europe. CFP has nine fully or partially owned refineries outside France, with marketing outlets in England, Belgium, Holland, Germany,

Switzerland, Italy, Austria, Greece, and Spain.[1] Outside Europe, CFP has refining and marketing operations in Africa, North America, South-East Asia, and Australasia.

SNEA sold about 30 per cent of its products outside France, primarily in Germany, Belgium, Holland, England, and Italy. It has one fully owned and one partially owned, refinery in Germany, plus smaller, all partially owned, refineries in Madagascar, Martinique, Cambodia, Senegal, Ivory Coast, and Gabon. In 1975, 81 per cent of SNEA's exploration investments were made outside France. In addition to oil, SNEA has interests in coal in Pennsylvania, Canada, South Africa, and Australia, and a 50 per cent share in a large nickel mining operation in New Caledonia.

Beginning in the early seventies, ENI began to advertise itself as a multinational company. By 1960, ENI had refining and/or marketing outlets in Germany, Austria, Switzerland, the Sudan, Morocco, Tunisia, Zaire, Tanzania, and Ghana. In 1969, ENI entered the Hungarian retail market, the first Western company to do so. About 18 per cent of ENI products were sold outside Italy. In 1978, ENI had six refineries outside Italy, in Germany, Tunisia, Ghana, Zaire, Tanzania, and Zambia. In addition to oil and gas, ENI was engaged in uranium mining ventures in Australia, Nigeria, Somalia, and Bolivia.

Newcomer STATOIL was already engaged in some marketing ventures in Denmark and Sweden through its partly owned company, NOROL. STATOIL also held a 50 per cent interest in the pipelines going from Ekofisk to Germany and England and a 5 per cent interest in the gas pipeline going from Frigg to the United Kingdom.

In a vertically integrated industry the aim is to control all phases of the operation and the necessity is to balance the different phases. As the sixties and seventies progressed, the four established NOCs became increasingly concerned with the balance among crude production, refining, and marketing. The NOCs were relative newcomers to the business; to pay the 'price of entry' they had had to make deals with the producing countries on terms more generous than those offered by the oil MNCs. Many of these deals included technical assistance, service contracts, and producer country participation – provisions more innovative than those in the concession agreements held by the oil MNCs. For NOC managements, these deals represented outstanding experiences in multinationalization. Keeping the balance among the phases of business in mind, CFP naturally chose to branch into refining and marketing outside France. ENI and SNEA naturally followed CFP's lead and established refining and marketing ventures outside Italy and France.

Multinationalization in downstream operations has several other economic consequences. First, multinationalization allows NOCs to market excess oil when their home markets become saturated. In France, for example, a saturated domestic market was a more important impetus to multinationalization than were the attractions of the foreign markets themselves. As it turned out for both CFP and SNEA, foreign downstream operations proved less profitable than their French operations. However, CFP, SNEA, and ENI were able to enter foreign downstream operations precisely because they had secure bases at home. A second consequence of multinationalization is that it allows the NOC to take advantage of economies of scale, which are very large in the oil industry. In both refining and marketing, investments must be huge to reduce the unit costs of production and distribution. With saturated and slowly growing home markets, NOCs enlarge their markets by spreading into foreign countries, so that they can run their big home country refineries at capacity. Third, multinationalization affects balances of payments. Until the early sixties Europe suffered from a 'dollar gap'. Multinationalization of production reduced the dollars needed to pay economic rents to the oil MNCs – particularly if the oil itself was domestic, as was the case with Franc Zone crude. Likewise, if downstream operations earned foreign exchange, either directly or indirectly by realizing economies of scale, such operations saved foreign exchange for France and Italy. BNOC and STATOIL are also making significant contributions to the British and Norwegian balances of payments.

Once under way, multinationalization of upstream and downstream operations had spiralling effects on the NOCs just as it had had on the MNCs. Since oil usually is discovered in large quantities, markets had to be found for it. Once markets were established, the search for yet more crude was intensified. SNEA offers a good illustration. When oil was discovered in Algeria – and the discovery was originally thought to be much larger than it actually turned out to be – SNEA immediately began to establish a market position in France and to seek downstream operations abroad as well. Once SNEA established positions domestically and internationally, it had to find more crude oil to serve these markets – looking for oil outside France, of course. Some of these exploration activities turned out to be extremely expensive. For instance, developing the medium-sized Frigg field in the Norwegian sector of the North Sea involved the minority holders, CFP and SNEA, in expenditures equivalent to those that developed the Concorde.[2] While part of the money for North Sea explorations came from domestic sources (French, German, and Italian), most of it was raised in the Eurobond and Eurocurrency markets – thus, of course, increasing the NOCs' multinationalization. Such financing also added to the conflicts between the NOCs and their home governments. Spending funds and foreign exchanges outside the home country may create a negative balance of payments as well as evoke criticism that the NOC is not spending enough at home to fulfill the governments' nonfinancial goals.

The conflict between the NOCs and their owner governments arises because the NOCs had been established originally to deal with specific economic, political, and social problems. One of the primary motivations was the desire to reduce the governments' dependence on the oil MNCs. The multinationals had controlled all phases of the oil industry in the major European countries. Moreover, all the oil MNCs were Anglo-Saxon, and five of the seven were Americans. Such foreign 'domination', initially an annoyance, became a consternation. NOC governments hoped that by establishing their 'own' companies they would increase their own international political and economic powers. Yet having established, assisted, and nurtured the NOCs, the owner-governments now find themselves in the position of the sorcerer: they cannot prevent their 'apprentice'

NOCs from joining the ranks of the oil MNCs. Originally, it was in the NOCs' best interests not to join the oil MNCs, because if they did, they could no longer offer a nonprivate, non-Anglo-Saxon, and thus a politically differentiated product. Yet the oil industry's geographic diversity, capital intensity, and available scale economies made multinationalization attractive, and the spreading of technical and political risk proved to be a necessity.

Multinationalization has also been spurred by the NOCs' organizational and managerial goals. As pointed out in Chapter 12, size is of prime importance in the oil industry, and many of the growth opportunities lie outside the NOCs' home countries. NOCs' managers also acquire prestige and power if they head operations with affiliates and subsidiaries abroad. NOC executives are enamored of power and prestige no less than their counterparts in the private sector. Managerial independence from the owners is, moreover, one of the most cherished goals of NOC managements. While several techniques can increase NOCs' managerial independence from their owner governments, none has proven more successful than multinationalization. Foreign affiliates cannot be controlled easily by owner governments, but they can by company executives. The temptation to become multinational appears irresistible – and it has not been resisted. In an interdependent world, it is a mark of success for any sizeable company – and all NOCs are such – to join the ranks of the multinational corporations.

Governments, of course, are aware of this movement and are fearful that they are losing control over their NOCs. Governments have also realized, albeit slowly, that 'in general, it is easier for a public agency to change behavior of a private organization than of another public agency.'[3] For instance, if a government wants its publicly owned tobacco industry to stay profitable, it is not likely to legislate an anti-smoking campaign; if a government wishes to run its national airline at a profit, it is not likely to require noise abatement devices, which are expensive and do not produce revenue. If governments truly wish their NOCs to become financially successful, and by and large they all do, they not only cannot prevent their NOCs from becoming multinational, but also must assist them to do so, if need be. NOCs will increasingly respond to the needs of the nations they service and the home governments' controls will erode correspondingly. The chief executive officers of the NOCs will probably continue to be members of the national economic and political establishment. But the executives and managers under them will be technocrats well versed in the international petroleum industry and in multinational management. As of 1980, the managers of CFP, ENI, SNEA, and STATOIL are people of exactly such backgrounds and interests. Similar executives are beginning to appear in BNOC's management, and even in Veba there are some multinational stirrings. The Seven Sisters are well on their way to becoming eleven, twelve, or thirteen.

References

1. *Total Information*, Paris, Spring 1979, pp. 17–25.
2. Private communication to the author, February 1977.
3. J. Q. Wilson and P. Rachel, 'Can The Government Regulate Itself?', *The Public Interest*, Winter 1977, pp. 4–14.

Index